Electronics Toolkit

Geoff Phillips

Newnes
An imprint of Butterworth-Heinemann
A division of Reed Educational and Professional Publishing Ltd
Linacre House, Jordan Hill, Oxford OX2 8DP

A member of the Reed Elsevier group

OXFORD LONDON BOSTON
MUNICH NEW DELHI SINGAPORE SYDNEY
TOKYO TORONTO WELLINGTON

First published 1993
Second Edition 1997

British Library Cataloguing in Publication Data
A catalogue record for this book is available from
the British Library

ISBN 0 7506 37900

Library of Congress Cataloguing in Publication Data
A catalogue record for this book is available from
the Library of Congress

Printed and bound in Great Britain by
Biddles Ltd, Guildford and King's Lynn

Contents

Preface

Electronics engineers involved in the design and development of electronic circuits are constantly referring to technical data. The speed and efficiency of the design and development process is usually hampered by the problem of locating the data quickly and then assimilating it. The objective of this book is to bring together fundamental facts, concepts and applications of electronic components and circuits and present them in a clear, concise and unambiguous format, resulting in a reference book that the engineer would rather have on his desk alongside his calculator or on his bench alongside his oscilloscope, rather than on the bookshelf. The engineer's job is not finished when the circuit diagram has been completed, the design for manufacture process is just as important if volume production is to be undertaken. The task of component selection and second or even third sourcing of alternative components is an essential part of the design process. This book attempts to assist the engineer with the component sourcing process by the inclusion of tables which present the electrical and mechanical parameters of electronic components which have evolved to be "industry standard" devices. The tables are in a format where the parameters of components from different manufacturers may be compared easily and a decision made as to their suitability for the engineer's circuit.

The presentation and layout of the data, circuit diagrams and text in this book was achieved by the use of the state-of-the-art computer applications. The text, tables and drawings were prepared by the author using Word for Windows by Microsoft. The circuit diagrams and component symbols were prepared by the author using an application called ISIS Illustrator by Labcenter Electronics.

1 Resistors

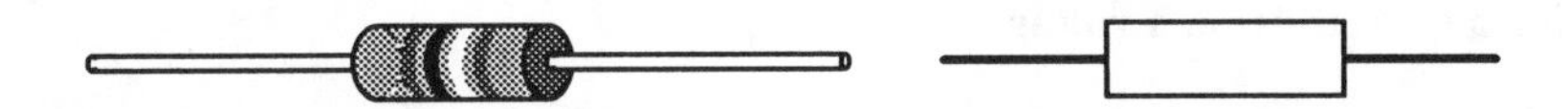

When a voltage is applied to a pure resistance, current flows in direct proportion to the voltage irrespective of how it may be varying.

The current through a resistor at any instant in time may be shown as:

$$i = \frac{v}{R}$$

v = the applied voltage at the same instant in time
R = the resistor's resistance in ohms

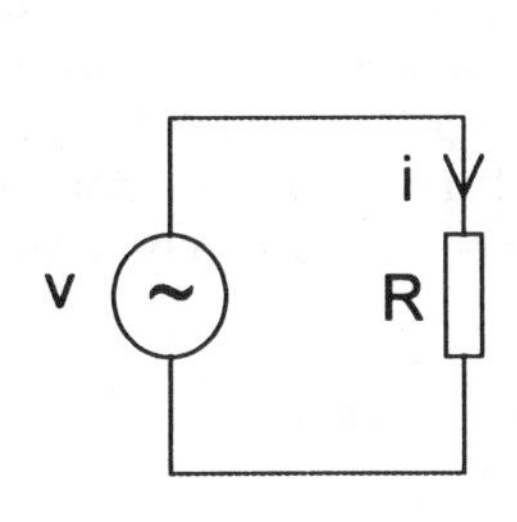

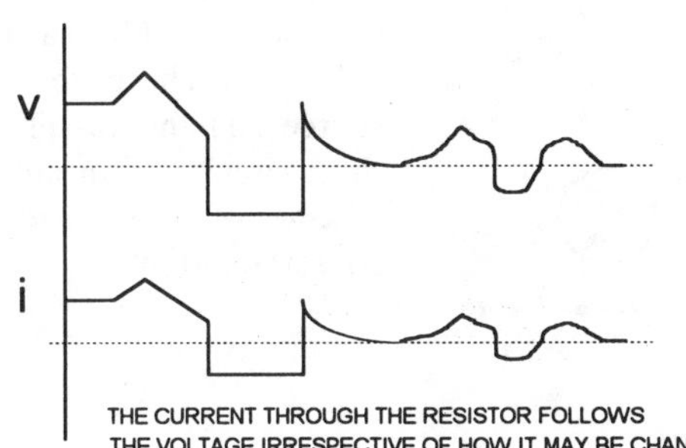

Power in a resistor

$$= I^2 R \quad \text{or} \quad = \frac{V^2}{R} \quad \text{WATTS}$$

Where I and V are r.m.s. or dc values of current and voltage respectively.

Rules of thumb

- When current is driven through a resistor, the input lead becomes positive.
- The total resistance of two resistors in parallel is product over sum (R1 x R2)/(R1 + R2).
- When analysing a complex circuit, ignore a resistor which is in parallel and > 10x.

Design pitfalls

Over-voltage

Always check that the resistor will not be subjected to a transient voltage which exceeds the manufacturer's maximum rating. An endurance test of the prototype may not reveal a problem but production units could fail.

Over-dissipation

If a 6W resistor is run at 6W or even 5W its surface temperature may be high enough to scorch printed circuit boards, damage surrounding components, or even melt solder. Always refer to the manufacturer's data sheet regarding surface temperature versus watts.

Pulsed power

Ensure that the resistor is capable of withstanding transient current surges.

Uses of a resistor

To derive a current from a voltage

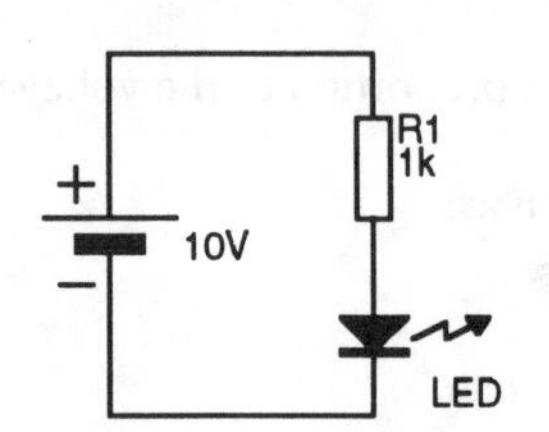

An LED cannot be connected directly to a dc supply as it is a low impedance and the current would be too high. A resistor may be used to give a current proportional to the applied voltage and in the example shown the resistor sets the current to approximately 10mA.

To derive a voltage from a current

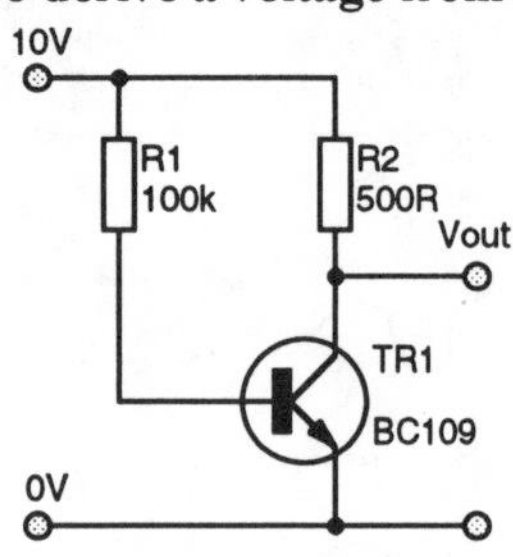

R1 derives a current of 0.1mA from the 10V supply to drive the base of TR1. TR1 has a gain of 100 and so 10mA flows in the collector. The collector of a transistor is a constant current source and the circuit would operate with R2 replaced by a short circuit. If an output voltage is required from the circuit however R2 is used to derive a voltage proportional to TR1's collector current.

Pull down

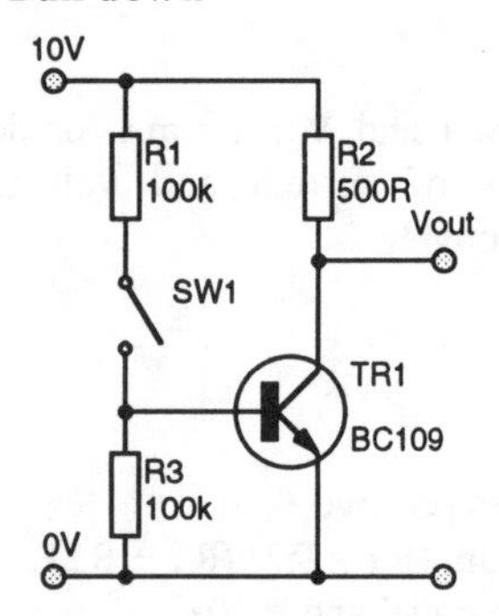

When the switch SW1 is opened, the base current to TR1 is interrupted and the transistor should switch off. There is leakage current across the collector-base junction however which flows back into the base and may cause the transistor to continue to conduct. Resistor R3 gives a pull-down effect to the base providing a path for the leakage current and assisting the turn-off of the transistor. R3 also improves the speed of turn-off in switching circuits.

Timing, filtering

See uses of a capacitor.

Potential division

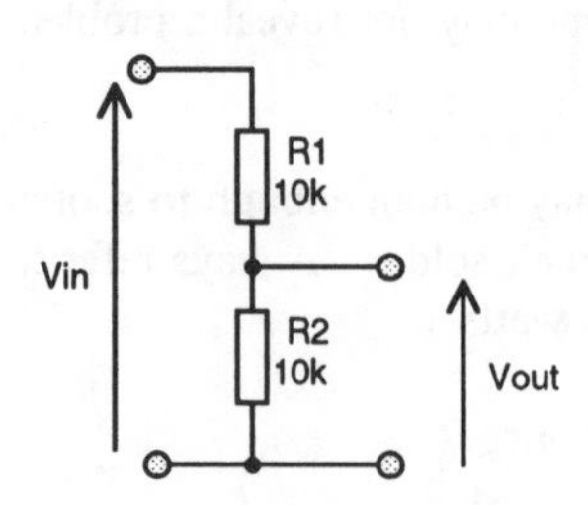

Two resistors in series constitute a potential divider which gives a voltage Vout equal to R2/(R1 + R2) times the input voltage.

$$Vout = Vin \frac{R2}{R1 + R2}$$

Current division

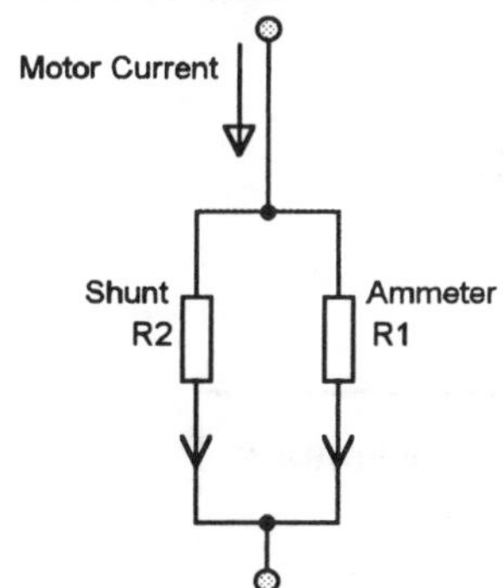

Two resistors in parallel may be used to split a current into two defined components according to the ratio of the resistor values. For example, the current drawn by a large electric motor has to be measured. The motor draws a maximum current of 100A but the ammeter which is to be used can only measure up to 1A full scale. A resistor may be placed in parallel with the ammeter so that one hundredth of the motor current passes through the ammeter and the remainder passes through the parallel resistor (often referred to as a shunt.) If the ammeter's resistance is R1 and the parallel resistor is R2 then the current through the ammeter is equal to R2 / (R1 + R2) times the motor current.

To derive a voltage proportional to a current

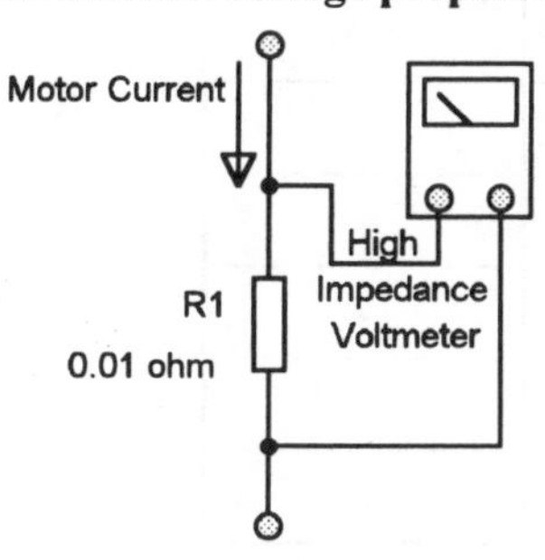

This application is similar to that on the previous page however the voltage derived in this case is used to measure the current. If a low value resistor is placed in series with the motor as shown in the previous example then a voltage will be produced across the resistor which is proportional to the current through it. The value of the resistor must be much lower than the resistance of the motor, that is to say the current drawn by the motor should not be significantly affected by placing the resistor in series. A high impedance voltmeter may then be used to measure the voltage across the resistor. The scale on the voltmeter may be actually calibrated in amps. In the example shown, a 0.01R resistor is used so when the motor draws 100A, 1V will be produced across the resistor. A 0 - 1V voltmeter may then be re-scaled to show 0-100A. The low value resistor is often referred to as a shunt but strictly speaking it does not "shunt" the current.

To discharge a capacitor

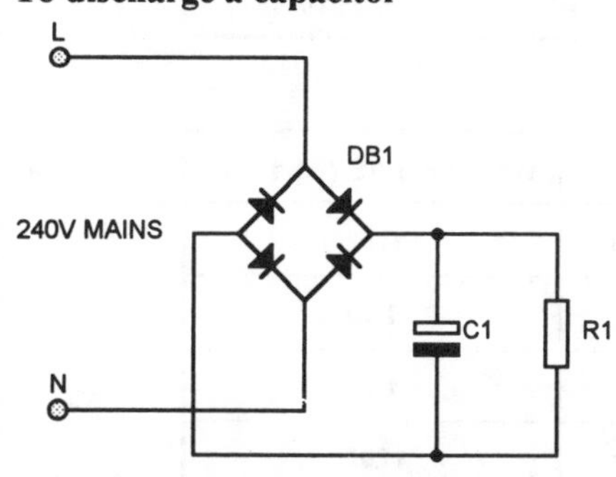

In most mains operated dc power supplies there is a reservoir capacitor which may be many thousands of microfarads. These capacitors can store charge for long periods of time after the mains has been switched off. A service engineer may inadvertently short circuit a reservoir capacitor while carrying out repairs. The resulting high current could burn out a printed circuit board copper track, not to mention the shock the engineer would get when a blinding flash emerges from the equipment. Manufacturers of equipment which incorporate high value electrolytic capacitors often fit a resistor across each capacitor to discharge it when the mains power is switched off. The value of the resistor must be able to discharge the capacitor in the time it takes for the engineer to remove all the screws to get inside the equipment. If the resistor is too low a value it may have to be a large wattage type and will waste power during operation of the equipment. In the example above, C1 is 1000μF. If R1 is 47k, then the voltage across C1 is discharged to a safe voltage in about two and a half minutes. During normal operation R1 dissipates approximately 2.5W.

Resistor colour codes

E12 and E24 series (4 band)

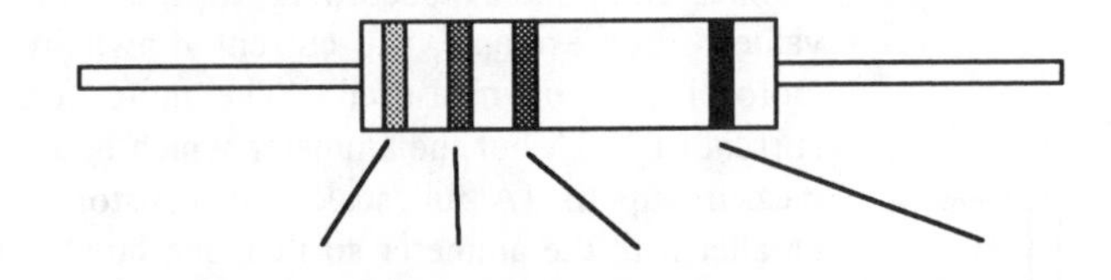

COLOUR	1st & 2nd DIGITS	MULTIPLIER	TOLERANCE
BLACK	0	1	-
BROWN	1	10	±1%
RED	2	10^2	±2%
ORANGE	3	10^3	-
YELLOW	4	10^4	-
GREEN	5	10^5	-
BLUE	6	10^6	-
VIOLET	7	10^7	-
GREY	8	10^8	-
WHITE	9	10^9	-
GOLD	-	10^{-1}	±5%
SILVER	-	10^{-2}	±10%

Examples

COLOUR	VALUE	WRITTEN AS (BS1852)
RED-RED-SILVER	0.22 ohms	R22
BROWN-BLACK-GOLD	1 ohm	1R0
BROWN-BLACK-BLACK	10 ohms	10R
BROWN-RED-BROWN	120 ohms	120R
BLUE-GREY-RED	6.8 kilohms	6K8
GREEN-BLUE-ORANGE	56 kilohms	56K
BROWN-RED-YELLOW	120 kilohms	120K
GREY-RED-GREEN	8.2 megohms	8M2

E48 and E96 series (5 band)

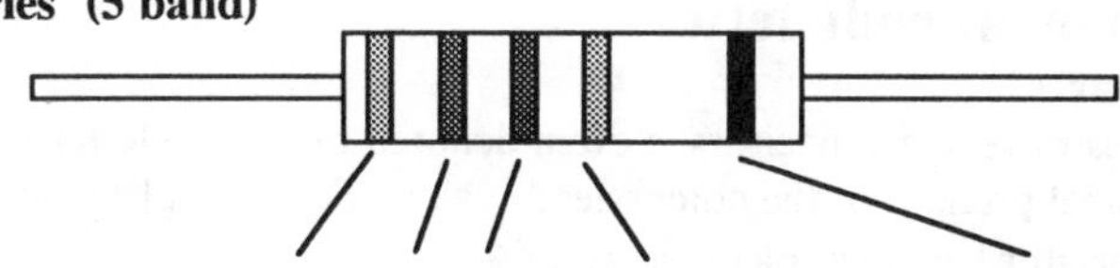

COLOUR	1st & 2nd & 3rd DIGITS	MULTIPLIER	TOLERANCE
BLACK	0	1	-
BROWN	1	10	±1%
RED	2	10^2	±2%
ORANGE	3	10^3	-
YELLOW	4	10^4	-
GREEN	5	10^5	-
BLUE	6	10^6	-
VIOLET	7	10^7	-
GREY	8	10^8	-
WHITE	9	10^9	-
GOLD	-	10^{-1}	±5%
SILVER	-	10^{-2}	±10%

Examples

COLOUR	VALUE	WRITTEN AS (BS1852)
RED-BROWN-GREEN-SILVER	2.15 ohms	2R15
ORANGE-ORANGE-RED-GOLD	33.2 ohm	33R2
GREEN-WHITE-BLACK-BLACK	590 ohms	590R
BROWN-RED-BROWN-BROWN	1.21 kilohms	1K21
ORANGE-ORANGE-RED-RED	33.2 kilohms	33K2
BROWN-RED-VIOLET-ORANGE	127 kilohms	127K
VIOLET-GREY-VIOLET-YELLOW	7.87 megohms	7M87
RED-RED-BLUE-GREEN	22.6 megohms	22M6

The BS1852 style of writing the resistor value avoids errors caused by the decimal point failing to reproduce clearly on drawings and photocopies. The letter R is used to signify ohms as the Greek letter Ω is not usually available on computer printers.

Passive component code letters

Tolerance code letters

The tolerance of resistors and capacitors is often denoted by a code letter which should not be confused with decimal prefixes of the component's value. For example a capacitor marked with 10K is a 10 μF capacitor having a tolerance of 10%.

CODE LETTER	TOLERANCE
A	±0.05%
B	±0.1%
C	±0.25%
D	±0.5%
F	±1%
G	±2%
J	±5%
K	±10%
M	±20%
N	±30%
Q	-10%, +30%
T	-10%, +50%
S	-20%, +50%
Z	-20%, +80%

Temperature coefficient code letters

Less common is a letter code to signify the component's drift in value with temperature.

CODE LETTER	TEMP COEFF
Y	15 ppm / °C
D	25 ppm / °C
C	50 ppm / °C
Z	100 ppm / °C
G	150 ppm / °C
X	250 ppm / °C

Derivation of E-resistor values

E12 Values

In the circuit below, the current flowing is 10A. Suppose we wish to reduce the current in equal steps. If resistor values of 2, 3, 4, 5, 6, 7, 8, 9, 10 ohms were chosen as standard values then the current would reduce in unequal steps.

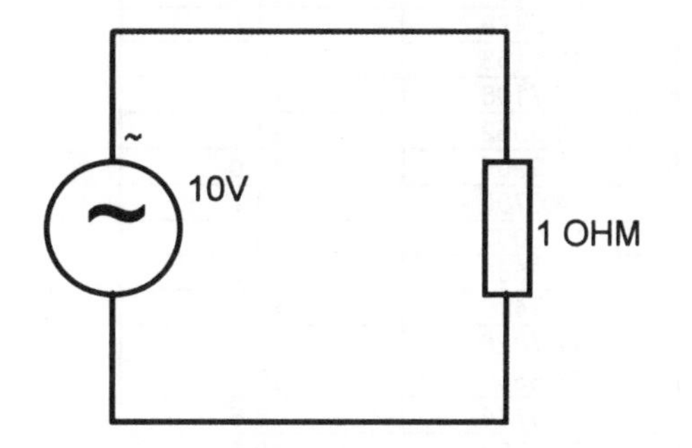

RESISTANCE	CURRENT
1 ohm	10.0A
2 ohm	5.00A
3 ohm	3.33A
4 ohm	2.50A
5 ohm	2.00A
6 ohm	1.67A
7 ohm	1.43A
8 ohm	1.25A
9 ohm	1.11A
10 ohm	1.00A

In the early days of resistor manufacture, tolerances of 10% were achievable. If a resistor of standard value 1 ohm was chosen then during manufacture the highest value allowable would be 1.1 ohm. It is reasonable to make the next highest standard value 20% higher on the reasoning that when it was at its minimum tolerance, then its value would overlap with 1.1 ohm (1.2 ohm x 0.9 = 1.08 ohm).

The 12th root of 10 is 1.2115 and this a convenient number of increments. If the standard values of resistors are set at 12 increments of x1.2115, then equal steps of reduction in current in the circuit above may be achieved.

EQUAL STEPS OF 12th ROOT OF 10	E12 VALUES ACTUALLY CHOSEN	CURRENT IN CIRCUIT	% REDUCTION IN CURRENT EACH STEP
1.0000	1	10.00	
1.2115	1.2	8.33	16.67
1.4678	1.5	6.67	20.00
1.7783	1.8	5.56	16.67
2.1544	2.2	4.55	18.18
2.6102	2.7	3.70	18.52
3.1623	3.3	3.03	18.18
3.8312	3.9	2.56	15.38
4.6416	4.7	2.13	17.02
5.6234	5.6	1.79	16.07
6.8129	6.8	1.47	17.65
8.2540	8.2	1.22	17.07
10.0000	10	1.00	18.00

The reader is left to decide why 2.7 was chosen instead of 2.6, 3.3 instead of 3.2 and 3.9 instead of 3.8. A similar anomaly exists in the choice of the preferred resistor values in the E24 range shown on the next page. For the E48 and E96 ranges however, the preferred values have been chosen as steps of the 48th and 96th root of 10, rounded to one decimal place.

E24 resistor values

24 EQUAL STEPS OF 24th ROOT OF 10	ACTUAL E24 VALUES CHOSEN
10.0000	**10**
11.0069	**11**
12.1153	**12**
13.3352	**13**
14.6780	**15**
16.1560	**16**
17.7828	**18**
19.5734	**20**
21.5444	**22**
23.7137	**24**
26.1016	**27**
28.7299	**30**
31.6228	**33**
34.8070	**36**
38.3119	**39**
42.1697	**43**
46.4159	**47**
51.0897	**51**
56.2342	**56**
61.8966	**62**
68.1292	**68**
74.9895	**75**
82.5405	**82**
90.8518	**91**
100.0000	**100**

E48 resistor values

48 EQUAL STEPS OF 48th ROOT OF 10	ACTUAL E48 VALUES CHOSEN
10	**10**
10.4914	**10.5**
11.0069	**11.0**
11.5478	**11.5**
12.1153	**12.1**
12.7106	**12.7**
13.3352	**13.3**
13.9905	**14.0**
14.6780	**14.7**
15.3993	**15.4**
16.1560	**16.2**
16.9499	**16.9**
17.7828	**17.8**
18.6566	**18.7**
19.5734	**19.6**
20.5352	**20.5**
21.5443	**21.5**
22.6030	**22.6**
23.7137	**23.7**
24.8790	**24.9**
26.1016	**26.1**
27.3842	**27.4**
28.7298	**28.7**
30.1416	**30.1**
31.6228	**31.6**
33.1767	**33.2**
34.8070	**34.8**
36.5174	**36.5**
38.3118	**38.3**
40.1945	**40.2**
42.1696	**42.2**
44.2418	**44.2**
46.4158	**46.4**
48.6967	**48.7**
51.0896	**51.1**
53.6002	**53.6**
56.2341	**56.2**
58.9974	**59.0**
61.8965	**61.9**
64.9381	**64.9**
68.1291	**68.1**
71.4770	**71.5**
74.9893	**75.0**
78.6743	**78.7**
82.5403	**82.5**
86.5963	**86.6**
90.8516	**90.9**
95.3161	**95.3**
100	**100**

96 STEPS OF 96th ROOT OF 10	E96 VALUES CHOSEN
10.0000	**10.0**
10.2428	**10.2**
10.4914	**10.5**
10.7461	**10.7**
11.0069	**11.0**
11.2741	**11.3**
11.5478	**11.5**
11.8281	**11.8**
12.1153	**12.1**
12.4094	**12.4**
12.7106	**12.7**
13.0192	**13.0**
13.3352	**13.3**
13.6589	**13.7**
13.9905	**14.0**
14.3301	**14.3**
14.6780	**14.7**
15.0343	**15.0**
15.3993	**15.4**
15.7731	**15.8**
16.1560	**16.2**
16.5482	**16.5**
16.9499	**16.9**
17.3613	**17.4**
17.7828	**17.8**
18.2145	**18.2**
18.6566	**18.7**
19.1095	**19.1**
19.5734	**19.6**
20.0486	**20.0**
20.5352	**20.5**
21.0337	**21.0**
21.5443	**21.5**
22.0673	**22.1**
22.6030	**22.6**
23.1517	**23.2**
23.7137	**23.7**
24.2894	**24.3**
24.8790	**24.9**
25.4829	**25.5**
26.1016	**26.1**
26.7352	**26.7**
27.3842	**27.4**
28.0489	**28.0**
28.7298	**28.7**
29.4272	**29.4**
30.1416	**30.1**
30.8733	**30.9**
31.6227	**31.6**

96 STEPS OF 96th ROOT OF 10	E96 VALUES CHOSEN
32.3904	**32.4**
33.1767	**33.2**
33.9820	**34.0**
34.8070	**34.8**
35.6519	**35.7**
36.5174	**36.5**
37.4038	**37.4**
38.3118	**38.3**
39.2419	**39.2**
40.1945	**40.2**
41.1702	**41.2**
42.1696	**42.2**
43.1933	**43.2**
44.2418	**44.2**
45.3158	**45.3**
46.4158	**46.4**
47.5426	**47.5**
48.6967	**48.7**
49.8788	**49.9**
51.0896	**51.1**
52.3298	**52.3**
53.6002	**53.6**
54.9013	**54.9**
56.2340	**56.2**
57.5991	**57.6**
58.9974	**59.0**
60.4295	**60.4**
61.8965	**61.9**
63.3990	**63.4**
64.9381	**64.9**
66.5144	**66.5**
68.1291	**68.1**
69.7829	**69.8**
71.4769	**71.5**
73.2121	**73.2**
74.9893	**75.0**
76.8097	**76.8**
78.6742	**78.7**
80.5841	**80.6**
82.5403	**82.5**
84.5439	**84.5**
86.5963	**86.6**
88.6984	**88.7**
90.8516	**90.9**
93.0570	**93.1**
95.3160	**95.3**
97.6298	**97.6**
100.0000	**100.0**

Resistor selection chart

TYPE	COMPOSITION	FILM	WIREWOUND	SMD
GENERAL ATTRIBUTES	Low inductance	Accurate tolerances available. Low drift and good temperature coefficient.	Higher power and pulsed power than film type.	Lower wattage Low inductance

RESISTOR TYPE	CARBON COMPOSITION	CARBON FILM	METAL FILM	METAL OXIDE	METAL GLAZE	ALUMINIUM CLAD	VITREOUS ENAMEL	CEMENTED	SMD CHIP
CON-STRUCTION	Moulded carbon	Homogeneous carbon film on ceramic body into which a helical groove is cut	Homogeneous metal alloy film on ceramic body into which a helical groove is cut	Homogeneous tin-oxide film on glass body into which a helical groove is cut	Metal-glazed film on a ceramic body	Wirewound housed in an aluminium case	Resistance wire is wound on a ceramic rod and coated with a vitreous enamel	Resistance wire is wound on a ceramic rod and coated with a silicon cement	Resistive paste applied to a ceramic monolith Laser-cut trimmed to value
ADVANTAGES	Low inductance Good pulse handling Fair high voltage performance	Low cost/performance ratio	High accuracy available Low temp drift	Good temperature coefficient, very low noise	High values High voltage	High power dissipation May be heat-sunk	High pulsed power capability	High pulsed power capability	Small physical size
DIS-ADVANTAGES	Poor tolerance Poor long term drift Poor temp coef High noise Poor availability	Modest temp drift	Higher cost	Higher cost	Limited value range	Higher inductance than film	Higher cost	Higher inductance than film	Limited wattage and voltage ratings
RESISTANCE RANGE AVAILABLE	10R to 22M	1R to 10M	5R to 10M	0R2 to 100k	100k to 22M	0R01 to 120k	0R1 to 100k	0R1 to 33k	1R to 10M
WATTAGES AVAILABLE	0.25W to 1W	0.125W to 2W	0.25W to 3W	0.5W TO 7W	0.25W to 1W	10W to 50W	2.5W to 12W	3W to 20W	0.25W (1206 style)
TEMPERATURE COEFFICIENT	+200 to -2000 ppm / °C	+200 to -1000 ppm / °C	±50 ppm / °C	±200 ppm / °C	±200 ppm / °C	±100 ppm / °C	+200 max ppm / °C	±100 ppm / °C	±200 ppm / °C
MANUFAC-TURERS	VTM Series BT Neohm CBT 1/2	Neohm CFR VTM DC Philips CR Piher PR	Philips MRS Neohm LR Welwyn GMF, MFR	Neohm ROX	Philips VR VTM GH	Arcol HS Welwyn WH	Welwyn W	ERG 74ER Philips AC03 - 20	Philips RC VTM RGC

1/4 W Carbon film resistors

MANFR	TYPE No	WATTAGE	DERATING	RANGE	TOL	TEMP COEF	MAX V	MAX TEMP	MAX DIMENSIONS
NEOHM	CFR25	0.25W @ 70°C	Linearly to 155°C	1R to 10M E24	±2%,±5%	0 to+800 R<10 -150to-800 R>10 ppm / °C	250V rms	+155°C	2.5 5.9 0.57
PHILIPS	CR25	0.33W @ 70°C	Linearly to 125°C	1R to 10M E24	±5%	-200 to -600 ppm / °C	250V rms	+155°C	2.5 6.5 0.6
PIHER	PR025	0.25W @ 70°C	Linearly to 155°C	1R to 10M	±2%,±5%	-150 to -800 ppm / °C	250V rms	+155°C	2.6 7.0 0.6
ROHM	R25X	0.33W @ 70°C	Linearly to 155°C	10R to 3M3	±2%,±5%	+300 to -1000 ppm / °C	300V rms	+155°C	2.5 6.6 0.63
WELWYN	GCF 1/4	0.25W @ 70°C	Linearly to 155°C	1R to 22M	±5%	-300 ppm / °C	250V rms	+155°C	2.5 6.4 0.6

1/2 W Carbon film resistors

MANFR	TYPE No	WATTAGE	DERATING	RANGE	TOL	TEMP COEF	MAX V	MAX TEMP	MAX DIMENSIONS
DRALORIC	LCA0411	0.5W @ 70°C	Linearly to 155°C	0R22 to 10M E24	±2%,±5%	-250 to -800 ppm / °C	350V rms	155°C	3.7, 10.5, 0.7
NEOHM	CFR50	0.5W @ 70°C	Linearly to 155°C	1R to 10M E24	±2%,±5%	0 to+800 R<10 -150to-850 R>10 ppm / °C	350V rms	155°C	3.3, 8.3, 0.74
PIHER	PR05	0.5W @ 70°C	Linearly to 155°C	1R to 10M	±2%,±5%	-130 to -800 ppm / °C	350V rms	155°C	3.7, 9.5, 0.6
ROEDER-STEIN	SK4	0.6W @ 70°C	Linearly to 125°C	1R to 30M	±5%	-300 to -1200 ppm / °C	500V rms	155°C	4.1, 12, 0.8
WELWYN	GCF 1/2	0.5W @ 70°C	Linearly to 155°C	1R to 2M2	±5%	-300 to -800 ppm / °C	350V rms	155°C	3.7, 9.0, 0.7

1W Carbon film resistors

MANFR	TYPE No	WATTAGE	DERATING	RANGE	TOL	TEMP COEF	MAX V	MAX TEMP	MAX DIMENSIONS
DRALORIC	LCA0719	1W @ 70°C	Linearly to 155°C	2R2 to 20M E24	±2%,±5%	-250 to -700 ppm / °C	750V rms	155°C	6.5 18.5 0.8
NEOHM	CFR100	1W @ 70°C	Linearly to 155°C	1R to 10M E24	±5%	0 to+800 R<10 -150to-900 R>10 ppm / °C	500V rms	155°C	5.2 13.3 0.82
PIHER	PR01	1W @ 70°C	Linearly to 125°C	1R to 10M	±2%,±5%	-150 to -800 ppm / °C	500V rms	155°C	5.0 13 0.8
ROEDER-STEIN	SK5	1.1W @ 70°C	Linearly to 125°C	1R to 24M E24	±5%	-200 to -1200 ppm / °C	750V rms	155°C	6 16 0.8

1/4 W Metal film resistors

MANFR	TYPE No	WATTAGE	DERATING	RANGE	TOL	TEMP COEF	MAX V	MAX TEMP	MAX DIMENSIONS
NEOHM	LR0204	0.25W @ 70°C	Linearly to 160°C	10R to 1M E24	±1%	±100 ppm / °C	200V rms	125°C	1.8 3.4 0.45
PHILIPS	MRS16T	0.4W@ 70°C	Linearly to 155°C	4R99 to 1M E24/E96	±1%	±50 ppm / °C	200V rms	125°C	1.9 3.7 0.5
PIHER	PM25	0.25W @ 70°C	Linearly to 150°C	10R to 1M	±1%	±50 ppm / °C	250V rms	155°C	2.3 6.5 0.6
ROEDER-STEIN	MK1	0.4W @ 70°C	Linearly to 175°C	0R22 to 5M1	±1,±2,±5%	±50 ppm / °C	250V rms	170°C	1.6 4.0 0.5
VTM	GP490	0.4W @ 70°C	Linearly to 155°C	1R to 5M1 E96	±1%	±50 ppm / °C	200V rms	155 °C	1.6 4.0 0.52
WELWYN	MFR3	0.4W @ 70°C	Linearly to 155°C	1R to 1M E96	±1%,±2%	±50 ppm / °C 10R to 150k	200V rms	155 °C	1.8 4.1 0.5

1/2 W metal film resistors

MANFR	TYPE No	WATTAGE	DERATING	RANGE	TOL	TEMP COEF	MAX V	MAX TEMP	MAX DIMENSIONS
DRALORIC	SMA0204	0.5W @ 70°C	Linearly to 155°C	1R to 10M E24/E96	±0.±1,±0.25, ±0.5,±1,±2,± 5%	±15,±25,±50, ±100 ppm / °C	200V rms	125°C	1.8 3.6 0.5
NEOHM	LR1	0.6W@ 70°C	Linearly to 155°C	1R to 10M E24	±1%, ±2% ±5% over 1M	±100,±50 ppm / °C	350V rms	125°C	2.5 5.9 0.6
PHILIPS	MRS25	0.6W @ 70°C	Linearly to 155°C	1R to 10M E24/E96	±1%	±50 to ±100 ppm / °C	350V rms	125°C	2.5 7.0 0.63
VTM	GP491	0.6W @ 70°C	Linearly to 155°C	1R to 10M E24/E96	±1%	±50 ppm / °C	350V rms	155 °C	2.5 6.0 0.62
WELWYN	MFR4	0.5W @ 70°C	Linearly to 155°C	1R to 1M	±0.5%,±1%, ±2%	±50 ppm / °C 10R to 150k	350V rms	155 °C	2.3 6.2 0.6

1/2 W metal oxide film resistors

MANFR	TYPE No	WATTAGE	DERATING	RANGE	TOL	TEMP COEF	MAX V	MAX TEMP	MAX DIMENSIONS
DRALORIC	SXA0411	0.7W @ 70°C	Linearly to 210°C	1R to 10M E24/E96	±1%	±50 to ±100 ppm / °C	350V rms	125°C	2.5 7.0 0.63
NEOHM	ROX05	0.5W @ 70°C	Linearly to 235°C	0R1 to 75k E24	±2%, ±5%	±300 ppm / °C	250V rms	235°C	3.5 10 0.7
WELWYN	MO 1/2 S	0.5W @ 70°C	Linearly to 235°C	10R to 50k E24	±5%,±10%	±350 ppm / °C	250V rms	155 °C	2.3 6.2 0.6

1/4 W metal glaze resistors

MANFR	TYPE No	WATTAGE	DERATING	RANGE	TOL	TEMP COEF	MAX V	MAX TEMP	MAX DIMENSIONS
PHILIPS	VR25	0.25W@ 70°C	Linearly to 155°C	100k to 22M	100k to 15M ±5% 15M to 22M ±10%	±200 ppm / °C	1600V dc 1150V rms	155°C	2.5 7.5 0.6
VTM	GH82	0.25W@ 70°C	Linearly to 155°C	100k to 22M E12,24,96	±1,±5,±10%	±200 ppm / °C	1600V dc 1150V rms	155 °C	2.5 6.0 0.62

1/2 W metal glaze resistors

MANFR	TYPE No	WATTAGE	DERATING	RANGE	TOL	TEMP COEF	MAX V	MAX TEMP	MAX DIMENSIONS
PHILIPS	VR37	0.5W @ 70°C	Linearly to 155°C	100k - 33M E24	±1%,E24/96 ±5%, E24	±200 ppm / °C	3.5kV dc 2.5kV rms	155°C	4.0 10 0.7
VTM	GH84	0.5W @ 70°C	Linearly to 155°C	100k to 33M E24,96	±1%,±5%	±200 ppm / °C	3.5kV dc 2.5kV rms	155 °C	4.0 9.0 0.7

1W metal glaze resistors

MANFR	TYPE No	WATTAGE	DERATING	RANGE	TOL	TEMP COEF	MAX V	MAX TEMP	MAX DIMENSIONS
PHILIPS	VR68	1W @ 70°C	Linearly to 155°C	100k to 68M E24,E96	±1%,E24/96 ±5%, E24	±200 ppm / °C	10kV dc 7kV rms	155°C	6.8 19 0.8
VTM	GH86	1W @ 70°C	Linearly to 155°C	100k to 68M E24,96	±1%,±5%	±200 ppm / °C	10kV dc 7kV rms	155 °C	6.8 16.5 0.7

7W ceramic case wire-wound resistors

MANFR	TYPE No	WATTAGE	TEMP RISE AT 7W	RANGE	TOL	TEMP COEF	MAX V	MAX DIMENSIONS
ISKRA	EIZ1107	7W at 70°C	230°C	0R1 to 15k	±5%, E24 ±10%,E12	<0.56R +1300 other -80 to +400 ppm / °C	Not quoted	6.5 6.5 38 0.82
MEC MICRON	MS	7W at 70°C	180°C	0R51 to 56k			Not quoted	10.5 10 36.5 0.81
NEOHM	RA7	7W at 70°C	200°C	0R1 to 2k2	±10%, ±5%	±20 to +250 ppm / °C	Square root of 7 x R	9.3 9.3 25.5 0.82
VTM	KH 210-8/0	7W @ 70°C	225°C	0R11 to 33k	±5%, E24 ±10%,E12	-80 to +500 ppm / °C	Square root of 7 x R	6.7 6.7 39 0.82

3W cement coated wire-wound resistors

MANFR	TYPE No	WATTAGE	TEMP RISE AT 3W	RANGE	TOL	TEMP COEF	MAX V	MAX TEMP	MAX DIMENSIONS
PHILIPS	AC03	3W at 40°C	275°C	0R1 to 3k	5%, E24	±600 R<10R -80 to +140 R>10R ppm / °C	1000V dc	350°C	5.5 13 0.8
ISKRA	EIZ1003	3W at 70°C	275°C	0R1 to 15k	±5%, E24 ±10%,E12	-80 to +1300 ppm / °C	Not quoted	205°C	5.0 34 0.82
VTM	CR256-0	3.5W@ 70°C	260°C	0R1 to 20k E24	±5%	±40 ppm / °C	Square root of 3.5 x R V rms	350 °C	5.5 17 0.8
WELWYN	WA84	2.5W @ 70°C	260°C	0R01 to2k2	±5,10%	>1R:200 <1R:350 ppm / °C	100V rms	275 °C	5.2 14.5 0.8

Variable resistors

Variable resistors are resistors which can be adjusted in value by some mechanical means. Examples include potentiometers, rheostats and trimmers, and are often used to make adjustments to electronic equipment such as volume control, sensitivity or gain. Strictly speaking a variable resistor is a two-terminal device whereas a potentiometer is a three-terminal device where a terminal called the wiper is able to move along the whole length of the element which constitutes the resistor and the other two terminals are connected to the ends of the resistive track.

Potentiometers

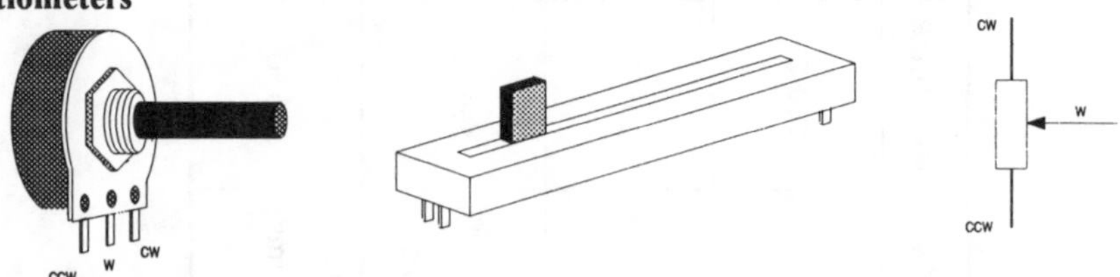

Potentiometers are available as the control type which are designed for continued use such as a volume control on an audio system. Both rotary and linear or slider types are available. Pre-set or trimmer potentiometers are generally designed for a limited number of operations and usually require a tool to operate. They are used to adjust an electronic circuit to a desired performance during manufacture or calibration.

Potentiometer law

This is the relationship between the mechanical movement of the wiper and the resistance between the wiper and the 0V terminal (usually the CCW terminal). The two main laws used by potentiometer manufacturers are linear and logarithmic. With a linear law potentiometer, the resistance varies linearly with movement of the wiper. Logarithmic law potentiometers are often used for volume controls in audio equipment as the human ear has an approximately logarithmic response to increases in loudness of sound.

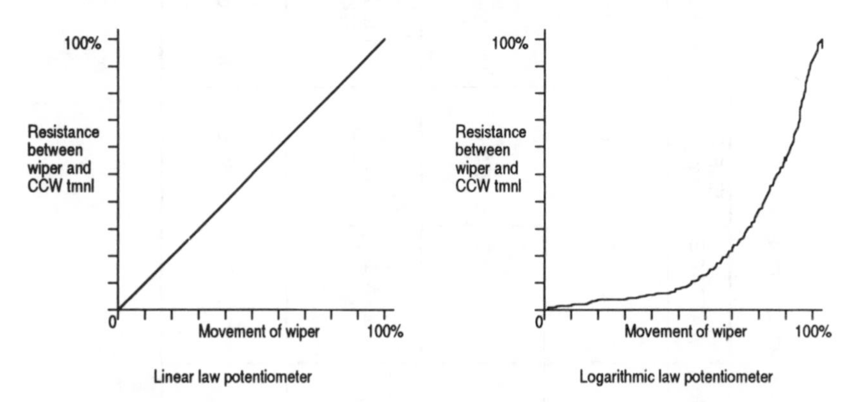

Linear law potentiometer

Logarithmic law potentiometer

Resistance track material

The material used for the potentiometer's resistance element or track may simply be a carbon film for low-cost general purpose use, a metal glaze track for high voltage applications such as a TV tube focus control, or a cermet track mainly used for higher quality, better temperature performance trimmer potentiometers. The professional audio industry often favours conductive plastic track slider potentiometers for their better tactile feel and lower noise performance.

Uses of a potentiometer

Volume control

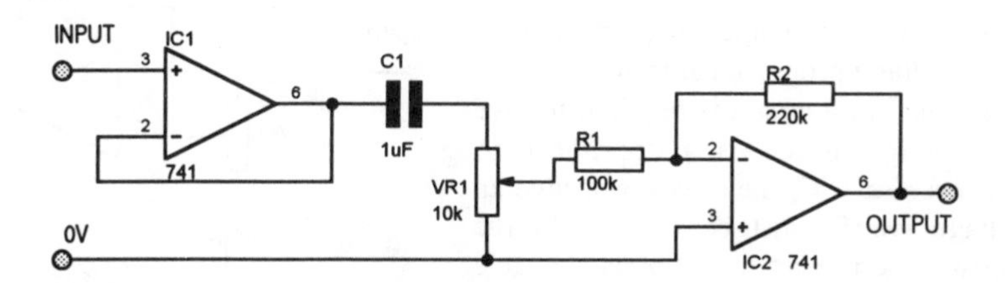

Logarithmic law potentiometers are usually used for volume controls in audio systems. The circuit preceding the volume control must be capable of driving the resistance of the potentiometer otherwise loss of signal will occur and the law of the potentiometer may be affected. Note the potentiometer is coupled to IC1 via a capacitor C1. If there is a dc content in the signal applied to the potentiometer, it may give a scratchy sound during operation. The dc content would also be amplified by the stage after the potentiometer which would cause the output voltage to shift thereby reducing the ac voltage headroom in one direction. C1 removes any dc component from the driving circuit. The reactance of C1 at the lowest audio frequency must be equal to or less than the potentiometer's resistance. The circuit which is connected to the wiper of the potentiometer must have a resistance greater than 10 times the resistance of the potentiometer otherwise loss of signal will occur and the law of the potentiometer may be affected.

Gain pre-set

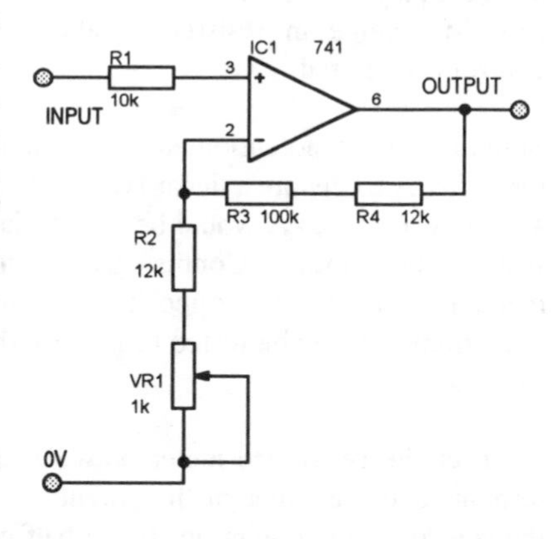

The circuit shown opposite is a non-inverting dc amplifier. The designer requires a gain of 10. A single resistor could be used as the feedback and a pre-set potentiometer connected between pin 2 of the op-amp and 0V. The potentiometer would then be adjusted until the gain was equal to 10. The user of the circuit would find the potentiometer difficult to set, however, because of its resolution. The circuit would have a poor stability with temperature and time due to the drift of the potentiometer. A fixed resistor should be used in series with a pre-set potentiometer such that the potentiometer's resistance is many times lower (ideally 10 times) than that of the resistor. The circuit is then easier to adjust accurately and will have a much better stability. The circuit shown uses two resistors in series as the feedback element to obtain a value which gives the required gain when the potentiometer is set approximately half way along its travel. The potentiometer is actually being used as a variable resistor in this application. When laying out the printed circuit board for this type of circuit it is good practice to arrange for clockwise rotation of the potentiometer to cause the gain to increase. The tolerance of the resistors must be taken into account when designing a circuit of this nature.

Offset or zero control

In dc amplifier applications, there may be dc offset voltages present due to imperfections in the op-amps or there may be an offset in the dc input signal. For example, a temperature sensor may produce a dc voltage proportional to temperature in degrees kelvin, but a circuit is required to give an output voltage which is proportional to degrees Celsius. At 0°C, the sensor will give an output voltage which must be offset. In the circuit opposite the pre-set potentiometer is able to provide an offset of ±3.66V at the output. The potentiometer is adjusted until the output voltage is zero when the sensor is at 0°C.

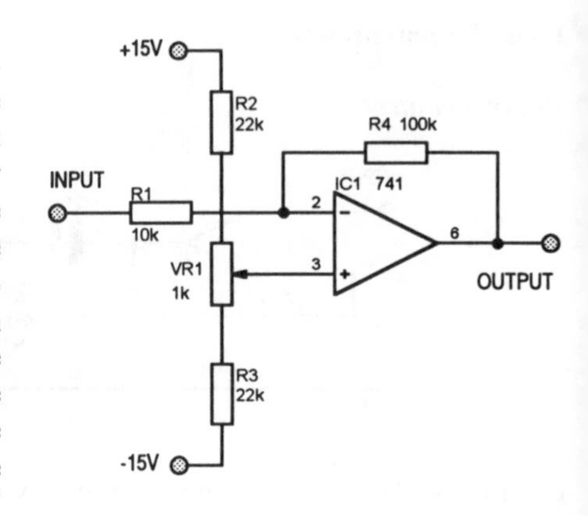

Testing a potentiometer

A resistance meter connected between the CW and CCW terminals of the potentiometer will confirm the resistance value printed on the body of the potentiometer. The value of the potentiometer may be determined in this way should the value be illegible. The design of slider potentiometers is not standardized and it may not be obvious to the engineer which pin is which. Pairs of potentiometer contacts may be tested whilst operating the slider control. The two end contacts may be identified as those which do not show a change in resistance value when the control is operated.

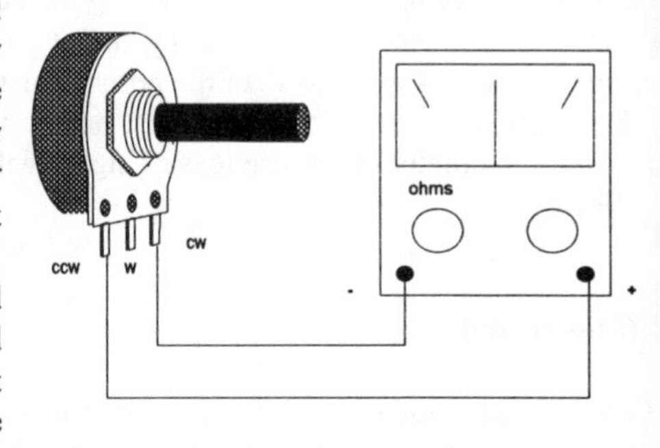

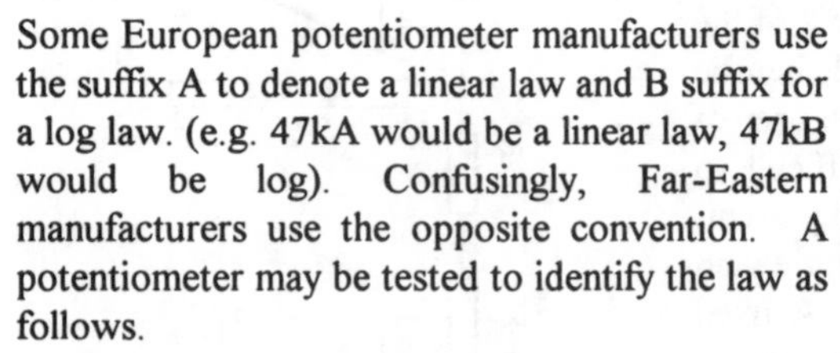

Some European potentiometer manufacturers use the suffix A to denote a linear law and B suffix for a log law. (e.g. 47kA would be a linear law, 47kB would be log). Confusingly, Far-Eastern manufacturers use the opposite convention. A potentiometer may be tested to identify the law as follows.

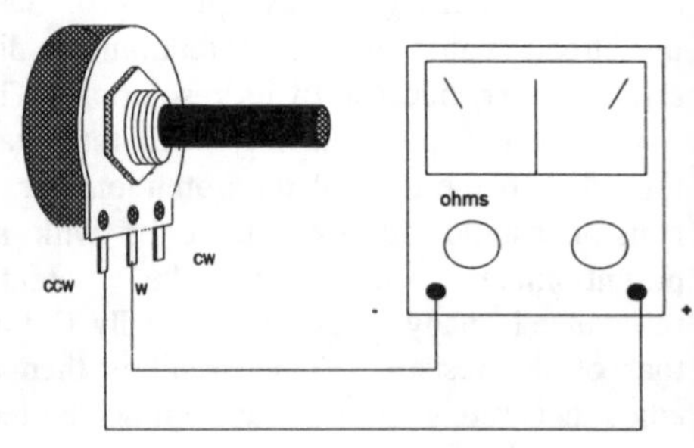

Connect the resistance meter between the CCW terminal and the wiper of the potentiometer. Set the control of the potentiometer to half way of its mechanical travel. If the resistance meter reads half the potentiometer's value then it is a linear law. If the resistance is approximately one tenth of the potentiometer's value then the law is logarithmic.

2 Capacitors

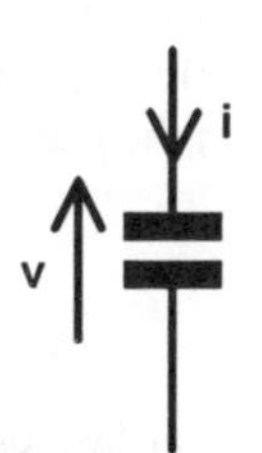

Electrical symbol

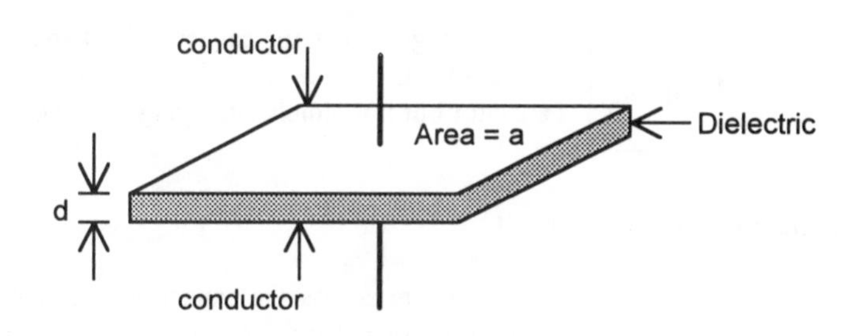

Parallel plate capacitor

Definition

Capacitance is the property which exists when two conductors are brought in close proximity to each other and are then able to store electrical charge.

The capacitance is increased if the area of the conductors is increased or the distance between them is reduced. The insulating material between the conductors is called the dielectric and also affects the value of capacitance.

Capacitance of a parallel plate capacitor

$$C = \frac{\varepsilon a}{d}$$

C = capacitance in Farads
a = area of one of the metal plates in square metres
d = distance between the plates in metres
ε = permittivity of the dielectric in Farads per metre

Charge stored in a capacitor

$$Q = CV$$

Q = stored charge in coulombs

The charge stored on a capacitor in coulombs is proportional to its capacitance and to the voltage to which it is charged. The measure of capacitance is how much charge it can store per volt. The charge stored is actually the charge difference between the plates. The positive plate gains a net charge of 1/2Q and the negative plate loses 1/2Q.

Energy stored in a capacitor

$$\text{Energy} = \frac{1}{2}CV^2$$

One may think that if Q coulombs are stored at a voltage V, then the energy available is Q x V= CV^2. However, this would only be true if the voltage remained at V as it would with a battery. The voltage across the capacitor falls as charge is removed. If the charge is removed at a constant current, then the voltage will fall linearly to zero and the average voltage across the capacitor will have been V/2. The energy removed is thus:

$$Q \times V/2 = CV \times V/2 = 1/2CV^2$$

Varying the voltage across a capacitor

$$i = C\frac{dV}{dt}$$

If the voltage across the capacitor is changing, then a current must flow so that Q = CV is maintained. The current appears to flow through the capacitor but it is simply supplying or removing the charge on the plates.

Capacitors in an ac circuit

$$X_C = \frac{1}{2\pi fC}$$

If the voltage across the capacitor is changing in a sinusoidal fashion, then the current that flows to and from the plates also has a sinusoidal waveform. The amplitude of the sinusoidal current which flows is proportional to the amplitude of the sinusoidal voltage across the capacitor. The current $I = V/X_C$ where X_C is the reactance of the capacitor in ohms.

Rules of thumb

- A capacitor stores charge and releases it when required.
- The voltage across a capacitor cannot change instantaneously.
- The current through an ideal capacitor can change instantaneously.
- In a sine wave ac circuit, the current leads the voltage by 90 degrees.
- A high value capacitor may sometimes be assumed to be a short circuit to ac.
- A capacitor blocks dc but allows ac to flow.
- A capacitor can release charge at a much higher current than that used to charge it, but cannot supply charge at a higher voltage than the voltage to which it was charged.

Enigma: disappearing energy

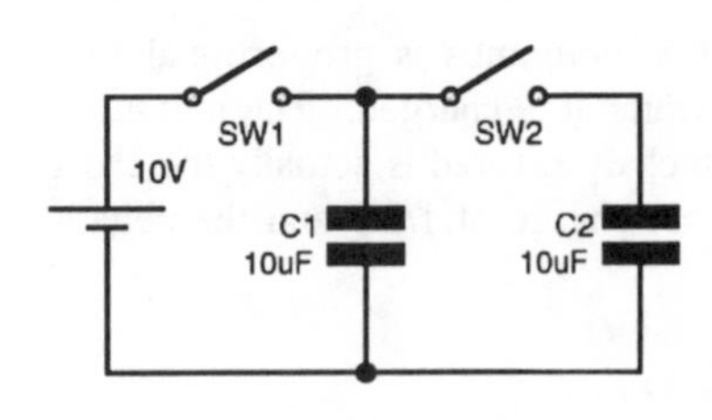

Charge and energy must be conserved. SW1 is closed and C1 is charged to 10V. The charge of the capacitor is then 100 microcoulombs and the energy is 500 microjoules. SW1 is then opened and SW2 closed. Charge must be conserved and the voltage must be the same across each capacitor, consequently the resulting voltage must be 10/2 = 5V as the resulting total capacitance is 20uF. The energy on the parallel combination of capacitors is thus $CV^2/2$ =250 microjoules. Where did the other 250 microjoules go?

Solution

The energy is actually dissipated in the resistance of the wire, switch and the losses of the capacitor. Whenever charge is moved from one place to another in this fashion, energy must be used. If 500 microjoules is to be stored on C1, then the battery has to supply 1000 microjoules. 500 are used to transfer the energy from the battery to C1. Energy may be transferred with minimum loss, however, via an inductor.

Equivalent circuit of a capacitor

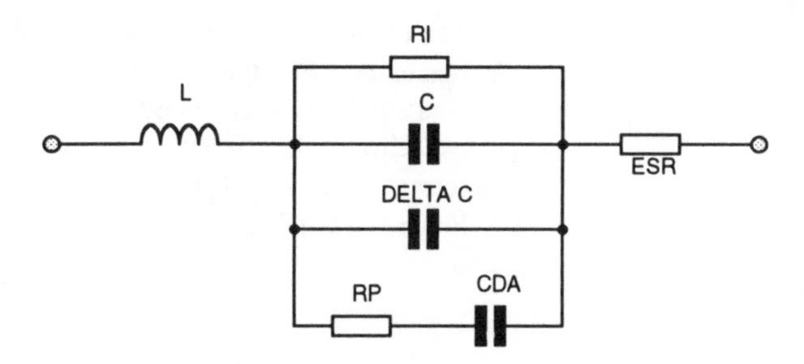

C = the nominal value of the capacitor
Delta C = the change in capacitance due to temperature, dc voltage or frequency
L = the inductance of the leads, metallization and the winding
ESR = the equivalent series resistance of the leads and the metallization
CDA = the dielectric absorption
RI = insulation resistance (of the dielectric)
RP = dielectric polarization resistance

Delta C
The value of a capacitor changes with changes in temperature and may be up to 10,000 parts per million per degree Celsius (1%). It may increase or decrease in value depending upon the type of capacitor.

L
The stray inductance of a capacitor causes it to be reduced in effectiveness at high frequencies. At the capacitor's resonant frequency, the reactance $X_L = 2\pi fL$ of the stray inductance is equal to the capacitive reactance $X_c = 1/2\pi fC$. With electrolytic capacitors, the resonant frequency may be in the order of 1MHz. With ceramic capacitors it is very much higher. If a capacitor is supplied with short leads it suggests that its stray inductance is even lower than the inductance of long leads.

ESR
The equivalent series resistance of a capacitor causes it to dissipate heat when the current becomes significant. This is why a maximum ripple current is quoted for electrolytic capacitors.

RI
The insulation resistance of a capacitor causes a leakage current to flow which may be a problem when the capacitor is used in a high impedance circuit such as a low frequency filter in the feedback circuit of an op-amp. Electrolytic capacitors have significant leakage currents.

CDA
Dielectric absorption is where charge is absorbed into the dielectric which cannot be removed as quickly as the charge stored on the plates of the capacitor due to the dielectric polarization resistance RP in series. The effect can be thought of as charge carriers penetrated into the body of the dielectric which have to flow through a very high resistance path in order to be extracted from the capacitor. The time constant of CDA and RP may be a few milliseconds.

Dissipation factor (tan δ)

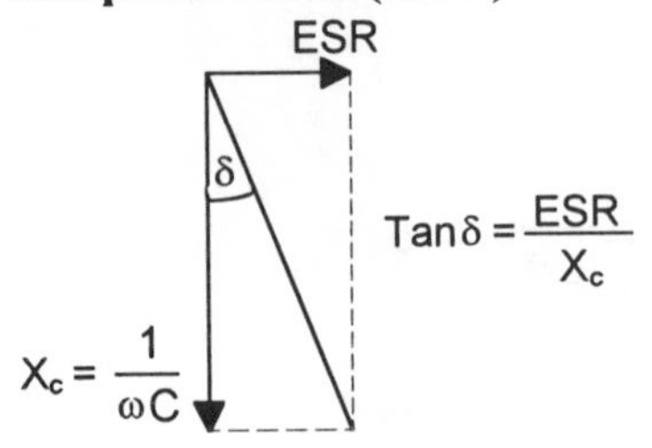

The dissipation factor is the ratio of the ESR to the reactance of the capacitor $1/2\pi fC$ ($=1/\omega C$) measured at a specified frequency. It enables capacitors of different types to be compared in terms of their self-heating characteristics. The lower the tan δ, the better.

dV/dt of a capacitor

This is a measure of a capacitor's ability to survive pulses of current. If the voltage across a capacitor is increased linearly from one value to another, then a current will flow to provide the charge required on the plates to satisfy the equation Q = C x voltage increase. The current is equal to C dV/dt. The metallized plates and the connections from the leads to the plates must be able to carry the current without rupturing. If the actual maximum pulse of current which a capacitor could withstand was quoted, a different value would have to be quoted for each value of capacitance. By quoting dV/dt, the capacitor manufacturers are able to state one value for a particular group of capacitors. The user may then calculate the maximum current a capacitor can withstand by i = C dV/dt.

Self-healing phenomenon

The plates of a metallized film capacitor are constructed by vacuum depositing a thin metal layer onto the dielectric film. If during the operation of a metallized film capacitor there is a breakdown of the dielectric film material due to an over-voltage or a manufacturing defect, a high fault current will flow through the "puncture" in the dielectric. This causes the metallized layer around the puncture to overheat and fuse open circuit. The puncture in the dielectric is effectively isolated from the remainder of the capacitor. This phenomenon is termed "self-healing". Each time a defect in the capacitor self-heals, the effective area of the plates is reduced; this results in a gradual reduction in capacitance over the life of the capacitor.

Uses of a capacitor

Energy storage or smoothing

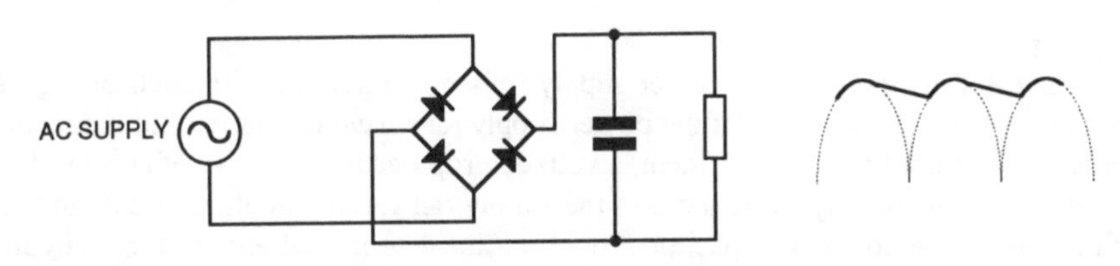

The capacitor receives energy during conduction of the diodes when the ac supply voltage is near the peak of its waveform. It releases the energy to the resistor when the diodes are not conducting and helps create a much smoother dc voltage across the resistor which is desirable for use as a dc power supply to electronic circuits. The capacitor will also help suppress over-voltage transients which may occur on the ac supply.

AC coupling and dc blocking

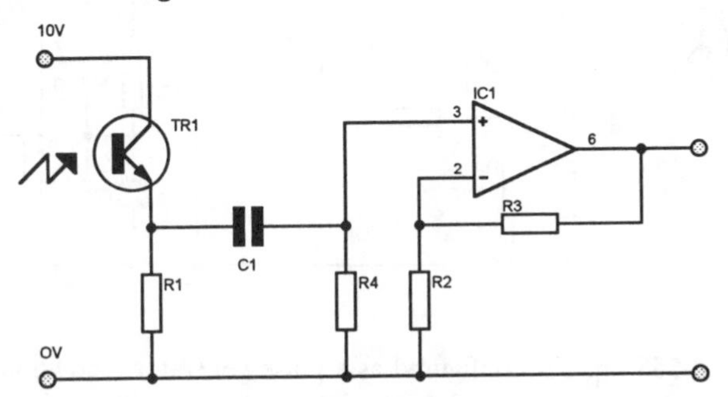

The capacitor provides a low impedance ac coupling from one circuit stage to another but blocks dc from flowing. In the circuit above, TR1 is an infra-red transistor which receives pulses of infra-red light from a TV remote control unit. A positive pulse of voltage appears at the emitter each time an infra-red pulse is received. There is also a dc voltage present at the emitter. The op-amp IC1 has been configured to amplify the voltage pulse. If the emitter of TR1 was connected directly to pin 3 of the op amp, it would amplify the dc also and the op amp would go into positive saturation. The capacitor C1 couples the pulse to IC1 but blocks the dc.

Rule of thumb

- The reactance of the capacitor at the lowest frequency it has to couple must be equal to the total resistance in series with the capacitor.

In order to find the total resistance in series with the capacitor:

1. Look at each capacitor lead in turn.
2. Determine how many paths there are from one lead of the capacitor.
3. Trace along each path and add up the resistances as you go along until you come to a low impedance point such as a dc rail or a low impedance output such as an op amp output.
4. Place the resistances of all the paths in parallel and calculate the total.
5. Repeat the procedure for the other capacitor lead.
6. Add the two totals.

The total resistance at the left hand lead of C1 is zero as the emitter of TR1 is a low impedance. The total resistance on the right hand lead is R4 as pin 3 of IC1 is a very high impedance.

De-coupling

Circuits connected to the same dc power supply rails may interfere with each other. This is because the copper wire or tracks of the power supply rails have resistance and inductance and when pulses of current flow through them, a voltage drop occurs. A pulse of current drawn by one circuit may consequently cause a dip in the supply rail feeding another circuit and cause it to malfunction. A de-coupling capacitor provides a local source of energy for a circuit stage which demands pulses of current and reduces the voltage dips in the dc supply rails thereby de-coupling that part of the circuit from the rest.

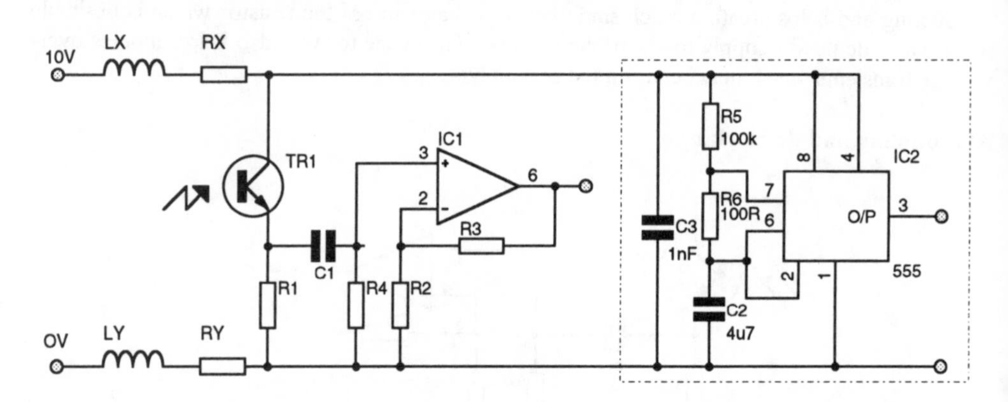

In the circuit above, a 555 timer is configured as a pulse generator (shown in the dotted box). It is connected to the same 10V dc rail as the infra-red high-gain receiver circuit. All the copper tracks of the printed circuit board will have resistance and inductance but the drawing shows Rx, Ry, Lx and Ly of the dc power supply tracks only. C3 de-couples the 555 circuit from IC1 circuit because it supplies the pulses of current required by IC2 with minimal dip in the 10V rail. If C3 was not present, the pulses of current would flow from the dc power supply and through Lx, Ly, Rx and Ry causing a transient voltage drop across them. The disturbance would appear across the supply to the IC1 circuit and would interfere with it.

Low-pass filter

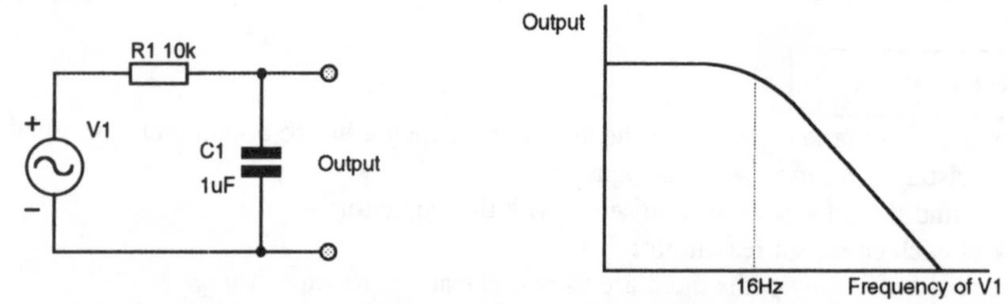

If a resistor is connected in series with a capacitor as shown, the combination acts as a low-pass filter having a cut-off frequency equal to $1/2\pi RC$. In the example above, the cut-off frequency is approximately 16Hz.

High-pass filter

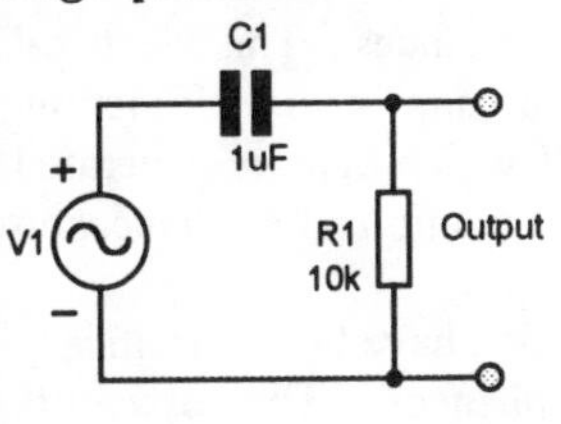

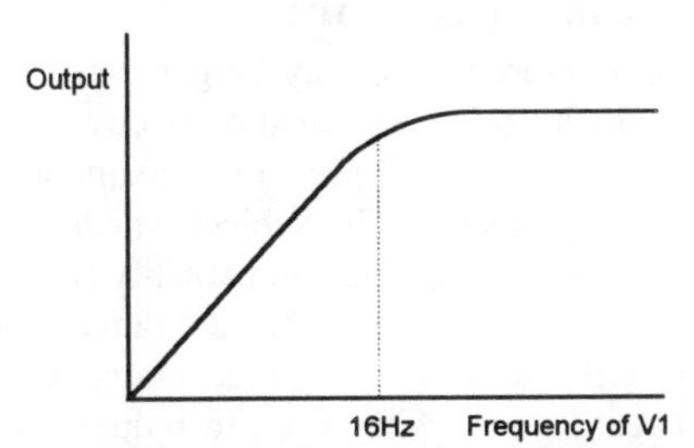

If a resistor is connected in series with a capacitor as shown, the combination acts as a high-pass filter having a cut-off frequency equal to $1/\ 2\pi RC$. In the example above, the cut-off frequency is approximately 16Hz.

Lossless voltage dropper

The circuit opposite gives a 9V dc supply from 240V ac mains. C1 is acting as a voltage dropper without dissipating energy, which would be the case if a resistor was used. R2 is included to limit the current to a safe value should the mains supply be switched on at its peak voltage. R1 discharges C1 when the mains is disconnected. The 9V supply is able to supply a few milliamps only. The circuit does not give isolation from the mains however and must not be used for applications where someone could touch the equipment being powered.

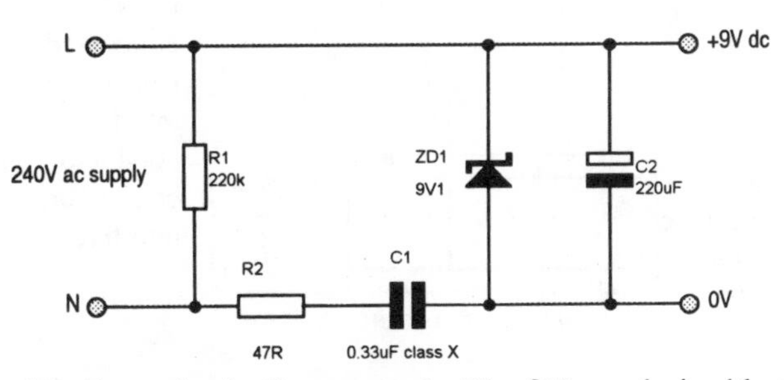

EMI suppression capacitors

EMI (electromagnetic interference) may be generated by electrical appliances. This electrical "noise" may interfere with other electrical or electronic equipment and cause it to malfunction. There are standards in force which state the maximum levels of EMI which may be generated by different types of appliances. The subject which considers the interaction of appliances in this respect is called electromagnetic compatibility (EMC) (see Chapter 6).

Capacitor manufacturers have introduced a range of capacitors which have been specifically designed to reduce the effects of EMI generated by electrical appliances. The capacitors considered in this section are mainly used to reduce the interference being transmitted via the mains supply.

Class X capacitors

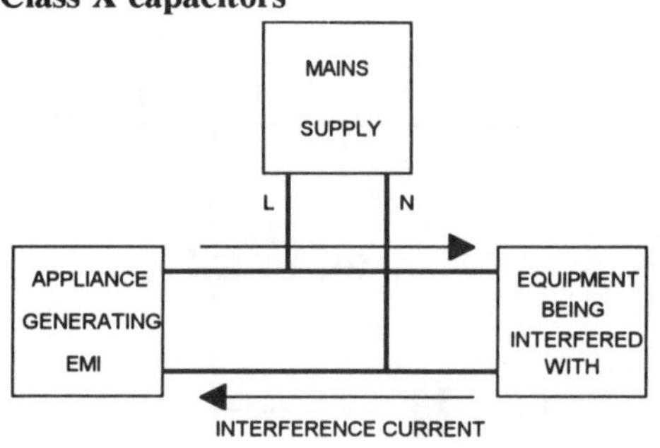

If a mains powered appliance is not earthed and is well separated from earthed equipment, then conducted EMI, generated by the appliance, can only interfere with other equipment by an interference current flowing down one mains conductor and returning via the other mains conductor. This is termed differential mode interference.

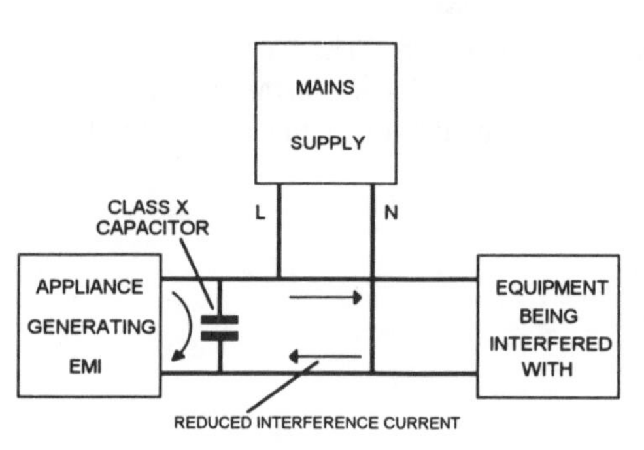

If a capacitor is connected across the mains supply at the appliance which is generating the EMI, then the differential mode interference current is offered an alternative path. Less of the interference current will therefore flow to the other equipment.

Capacitors which are designed to be applied in this way are termed class X capacitors. Class X capacitors are suitable for use in situations where failure of the capacitor would not lead to danger of electric shock to anyone using the appliance.

There are two classes of X capacitors, X1 and X2.

Class X1 (> 1.2kV)

High grade capacitors which are designed to be used in applications where transient voltages may occur on the mains supply which may exceed 1.2kV peak.

Class X2 (< 1.2kV)

General purpose grade capacitors which are designed to be used in applications where transient voltages will be less than or equal to 1.2kV peak.

Mnemonics

Class X capacitors are across (a cross X) the mains supply.
Class X1 is No.1 grade therefore > 1.2kV (1 is higher ranking than 2).

Class Y capacitors
If an appliance has a metal case which is earthed or has metal parts which are earthed, then there will be stray capacitances between the electrical circuitry and the earthed metal parts. Interference currents can then flow through the stray capacitances, travel along the earth conductor to other equipment, flow through the stray capacitances of that equipment and return to the appliance via the live and neutral conductors. This type of conducted interference is termed common-mode interference.

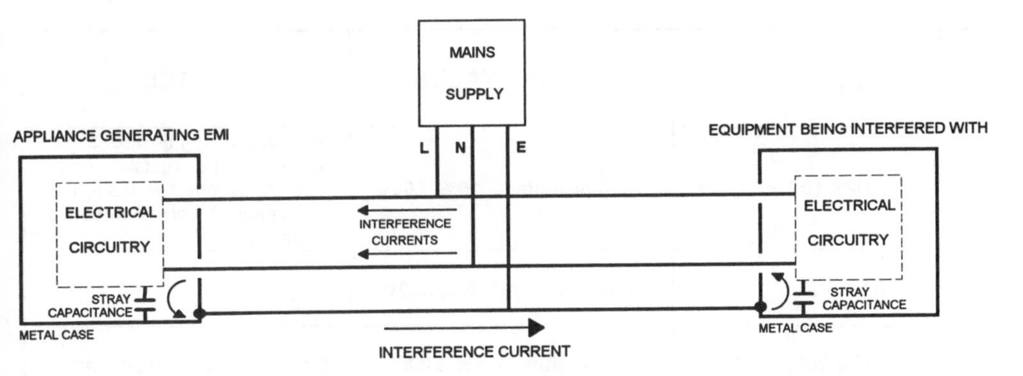

If two capacitors are connected, one between the live and earth, and the other between neutral and earth at the appliance, then the common-mode interference currents are offered an alternative path. Less of the common-mode interference currents will flow to the other equipment.

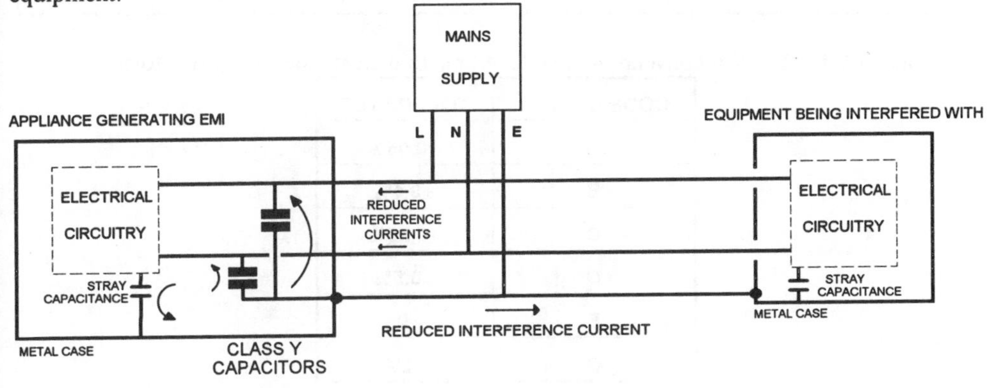

If the earth wire to the metal case of the appliance became disconnected, however, then one of the Y capacitors forms a direct connection between live and the metal case. If someone touched the metal case of the appliance, they may complete a circuit to earth. The value of the Y capacitor must be such that the maximum leakage current which will flow to earth is well below the value which may cause electric shock. A figure standardized for household equipment and portable tools is <0.5mA. If the Y capacitor failed to short circuit when the earth wire is disconnected, the metal case of the appliance would be directly connected to the live conductor and would expose the user to electric shock. It is for this reason that the electrical and mechanical safety margins are increased in the design of class Y capacitors. Manufacturers are very careful not to state that the capacitor will never fail to short circuit, however the inference is that the probability that it will fail to short circuit is remote.

Capacitor value marking systems

Some capacitor manufacturers make life easy and mark the value as it would be written i.e.:

10uF , 5%, 100V

European manufacturers tend to use a comma instead of the decimal point i.e. 3.3uF would be written **3,3uF**

Most manufacturers seem to have adopted the following convention: value first, then tolerance, then voltage. The following table gives examples of the different methods of marking.

MARKING	ACTUAL VALUE	NOTES
102K 100	1000 picofarads, 10%, 100V	The first 3 digits are in the format of the resistor colour code i.e. 10 and two 0's = 1000 the value being number of picofarads
0.01/5/250	0.01 microfarad, 5%, 250V	
4n7 J 2000	4.7 nanofarads, 5%, 2000V	Equivalent to 4700 picofarads
0,47 J 100	0.47 microfarads, 5%, 100V	
1.0K100	1 microfarad, 10%, 100V	

Some manufacturers use a convention of code letters to indicate the capacitor's tolerance.

CODE LETTER	TOLERANCE
A	±0.05%
B	±0.1%
C	±0.25%
D	±0.5%
F	±1%
G	±2%
J	±5%
K	±10%
M	±20%
N	±30%
Q	-10%, +30%
T	-10%, +50%
S	-20%, +50%
Z	-20%, +80%

Capacitor selection chart

CAPACITOR TYPE	CERAMIC NPO (COG)	CERAMIC X7R	CERAMIC Z5U	SILVER MICA	ALUMINIUM ELECTROLYTIC	TANTALUM
CON-STRUCTION	Thin metal layer deposited on ceramic plates	Thin metal layer deposited on ceramic plates	Thin metal layer deposited on ceramic plates	Layer of silver is applied to either side of thin mica sheet	The dielectric is an oxide layer which formed on the aluminium plates by electrochemical action	Porous tantalum pellet with electrolyte filling
ADVANTAGES	Very low temperature coefficient		High capacitance for medium size. Low cost	Very high stability Very high precision Good pulse handling	High capacitance for small physical size Low cost	High capacitance for small physical size Low cost. Lower leakage current than aluminium
DIS-ADVANTAGES	Limited capacitance range Higher values are high cost		Poor temperature coefficient	High cost Large size	Limited lifetime especially at high temperature High leakage current	Limited lifetime especially at high temperature Higher cost than aluminium
CAPACITANCE RANGE AVAILABLE	1.8pF to 0.47uF	10pF to 1uF	10pF to 2.2uF	2.2pF to 0.01uF	1uF to 470,000uF	0.1uF to 100uF
VOLTAGES AVAILABLE	50 to 200V	50 to 200V	50 to 100V	100 to 400V	6.3V to 400V	6.3V to 50V
TOLERANCES AVAILABLE	±5%, ±10%	±10%, ±20%	±10%, ±20%	±1%	±20%, -10 +20%	±10%, 20%
TEMPERATURE COEFFICIENT	0 to ±30 ppm / °C	±1000 ppm / °C	±10,000 ppm / °C	-20 to +100 ppm / °C	+1000 to +10,000 ppm / °C	+1000 ppm / °C
MANUFAC-TURERS	Kemet, Philips, Thomson, KVX - Kyrocera, Panasonic, Siemens	Philips, Thomson, KVX - Kyrocera, Panasonic, Siemens	Philips, Siemens, Kemet	Ashcroft, Iskra, Wimpey-Dubilier	Philips, Siemens, Wimpey-Dubilier	Kemet, Wimpey-Dubilier

For film capacitors, see next page

Film capacitor selection chart

TYPE	FILM / FOIL	METALLIZED FILM
CONSTRUCTION	Two aluminium foil electrodes separated by a dielectric of plastic film material	Thin metal layer vacuum deposited onto a dielectric film or a carrier film
ADVANTAGES	High insulation resistance. Good capacitance stability. Good current-carrying ability	Small physical size. Self-healing ability
DIS-ADVANTAGES	Large physical dimensions	Lower voltage withstand capability than film/foil Lower current carrying capability than film/foil

CAPACITOR TYPE	POLYCARBONATE KC	POLYESTER KT	POLYPROPYLENE KP	POLYSTYRENE KS	METALLIZED POLYCARBONATE MKC	METALLIZED POLYESTER MKT	METALLIZED POLYPROPYLENE MKP
ADVANTAGES	Good long-term stability Low TC	High voltages available Good pulse handling Good long-term stability	Low diss factor Excellent pulse perf Low losses Low dielectric absorp.	Extremely low losses Low dielectric absorp. Low -ve TC	Low TC Low cost	Good self-healing	Low losses Low dielectric absorp. High insulation R
DIS-ADVANTAGES	Higher cost	Poor TC	Higher cost Larger size	Poor soldering perf Affected by some solvents	Higher cost	Higher cost	Higher cost Larger size
APPLICATIONS	General purpose Time delay	Bypassing, coupling	Pulse applications SMPS	Timers, filters, sample and hold	General purpose	-	Sample/hold circuits Pulse applications SMPS, High f decoupling
TEMPERATURE COEFFICIENT	+150 ppm / ^{o}C	+500 ppm / ^{o}C	-200 ppm / ^{o}C	-150ppm / ^{o}C	+150ppm / ^{o}C or ±100ppm / ^{o}C	+400 ppm / ^{o}C	-200 ppm / ^{o}C
MANUFACTURERS	Roederstein KC1850 Rifa-Evox CFR Wima FKC2 & 3	Wima FKS2	Roederstein KP1830 Rifa-Evox PFR Wima FKP2 Philips 460,1,2,3,4 LCR PC/HV/S	Rifa-Evox SFA,SFR,SFE Suflex HS Philips 424,5,6,7	Philips 344 Rifa-Evox CMK Wima FKC2 Roederstein MKC-1858,60,62	Roederstein MKT Rifa-Evox MMK Siemens B32560 Philips - 365,8, 370,1,2,3 Wima MKS4 Thomson BF	Philips 378 Rifa-Evox PMR Roederstein-MKP1837,39,40,41,45,46

5mm lead pitch - polycarbonate film and foil capacitors

KC

MANFR	TYPE No	C VALUE	STD TOL	MAX V	MAX dV/dt	INSUL R	Tan δ	TEMP CO (0-80°C)	TEMP RANGE	MAX DIMENSIONS
EVOX	CFR5	0.01μF	±10%, ±5%,±2.5%	100V dc 63V ac	1000V/μs	>100 G	0.4% max at 10kHz	-25 ppm / °C	-55 to +125 °C	⌀ 0.5, 5, 9.0, 4.5, 7.2
ROEDER-STEIN	KC1850	0.01μF	±20%, ±10%, ±5%	63V dc 40V ac	1000V/μs	>500G	0.4% max at 10kHz	+25 ppm / °C	-55 to +100 °C	⌀ 0.5, 5, 8.5, 3.5, 7.2
WIMA	FKC2	0.01μF	±10%, ±5%,±2.5%	100V dc 63V ac	1000V/μs	>500G	0.4% max at 10kHz	-125 ppm / °C	-55 to +100 °C	⌀ 0.5, 5, 9.5, 4.5, 7.2

5mm lead pitch - polypropylene film and foil capacitors **KP**

MANFR	TYPE No	C VALUE	STD TOL	MAX V	MAX dV/dt	INSUL R	Tan δ	TEMP CO (0-80°C)	TEMP RANGE	MAX DIMENSIONS
EVOX	PFR5	0.01μF	±10%, ±5%,2.5% ±2%,±1%	63V dc 40V ac	1000V/μs	>100 G	0.07% max at 10kHz	-200 ppm / °C	-55 to +100 °C	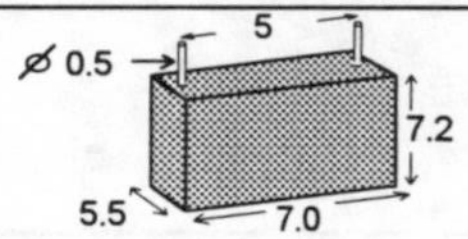 ⌀ 0.5, 5, 7.2, 5.5, 7.0
ROEDER-STEIN	KP1830	0.01μF	±10%, ±5%,2.5%, ±1%	63V dc 40V ac	1000V/μs	>500G	0.04% max at 10kHz	-125 ppm / °C	-55 to +100 °C	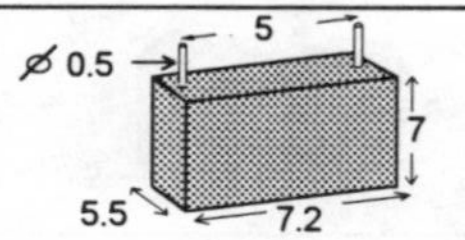 ⌀ 0.5, 5, 7, 5.5, 7.2
WIMA	FKP2	0.01μF	±10%, 5%,±2.5%,	63V dc 40V ac	1000V/μs	>500G	0.04% max at 10kHz	-375 ppm / °C	-55 to +85 °C	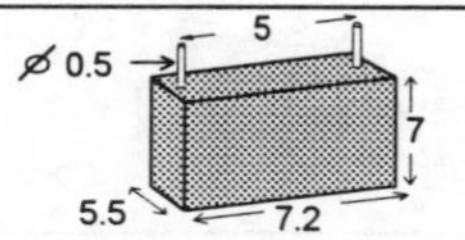 ⌀ 0.5, 5, 7, 5.5, 7.2

5 mm lead pitch - metallized polycarbonate capacitors **MKC**

MANFR	TYPE No	C VALUE	STD TOL	MAX V	MAX dV/dt	INSUL R	Tan δ	TEMP CO (0-80°C)	TEMP RANGE	MAX DIMENSIONS
EVOX	CMK5	0.1μF	±20%, ±10%, ±5%,±2.5%	63V dc 40V ac	10V/μs	>15 G	0.5% max at 10kHz	-12.5 ppm / °C	-55 to +125 °C	⌀ 0.5, 5, 8.0, 3.5, 7.2
ROEDER-STEIN	MKC1858	0.1μF	±20%, ±10%, ±5%	63V dc 40V ac	17V/μs	>3.75G	0.5% max at 10kHz	-25 ppm / °C	-55 to +100 °C	⌀ 0.5, 5, 8.5, 3.5, 7.5
WIMA	MKC2	0.1μF	±20%, ±10%, ±5%	63V dc 40V ac	15V/μs	>3.75G	0.5% max at 10kHz	-88 ppm / °C	-55 to +100 °C	⌀ 0.5, 5, 8.5, 3.5, 7.2

5 mm lead pitch - metallized polyester capacitors

MKT

MANFR	TYPE No	C VALUE	STD TOL	MAX V	MAX dV/dt	INSUL R	Tan δ	TEMP CO (0-80°C)	TEMP RANGE	MAX DIMENSIONS
EVOX	MMK5	0.01μF	±20%, ±10%,±5%,	250V dc 160V ac	40V/μs	>30 G	1.2% max at 10kHz	+375 ppm / °C	-55 to +100 °C	Ø 0.5, 5, 6.5, 7.2, 2.5
PHILIPS	370	0.01μF	±20%, ±10%,±5%,	250V dc	330V/μs	>15 G	1.3% at 10kHz	+375 ppm / °C	-55 to +85 °C	Ø 0.6, 5.08, 6.5, 7.2, 2.5
ROEDER-STEIN	MKT1817	0.01μF	±20%, ±10%,±5%,	250V dc 160V ac	44V/μs	>7.5 G	1.5% at 10kHz	+375 ppm / °C	-55 to +100 °C	Ø 0.5, 5, 6.0, 7.5, 2.5
SIEMENS	B32529	0.01μF	±10%	250V dc 160V ac	250V/μs	>7.5 G	1.5% at 10kHz	+500 ppm / °C	-55 to +100 °C	Ø 0.5, 5, 6.5, 7.2, 2.5
THOMSON	BF01	0.01μF	±20%, ±10%,±5%,	250V dc 160V ac	110V/μs	>7.5 G	1% at 10kHz	+400 ppm / °C	-55 to +100 °C	Ø 0.6, 5.08, 6.5, 7.5, 2.5
WIMA	MKS2	0.01μF	±20%,±10%	250V dc 160V ac	50V/μs	-	1.5% at 10kHz	+375 ppm / °C	-55 to +100 °C	Ø 0.5, 5, 7.0, 7.2, 2.5

Metallized polyester capacitors

MKT

MANFR	TYPE No	C VALUE	STD TOL	MAX V	MAX dV/dt	INSUL R	Tan δ	TEMP CO (0-80°C)	TEMP RANGE	MAX DIMENSIONS
EVOX	MMK22.5	1μF	±20%, ±10%,±5%,	250V dc 160V ac	8V/μs	>5 G	1.5% max at 10kHz	+375 ppm / °C	-55 to +100 °C	⌀ 0.8, 22.5, 16, 8.0, 26
PHILIPS	373	1μF	±10%	250V dc 160V ac	16V/μs	>10 G	1.5% max at 10kHz	+350 ppm / °C	-55 to +85 °C	⌀ 0.8, 15, 15, 8.5, 17.5
ROEDER-STEIN	MKT1822	1μF	±20%, ±10%,±5%,	250V dc 160V ac	8V/μs	>10 G	1.5% at 10kHz	+375 ppm / °C	-55 to +100 °C	⌀ 0.8, 22.5, 16.5, 8.5, 26.5
SIEMENS	B32523	1μF	±10%,±5%	250V dc 160V ac	4V/μs	>2.5 G	1.0% at 1kHz	+500 ppm / °C	-55 to +100 °C	⌀ 0.8, 22.5, 16.5, 8.5, 26.5
THOMSON	BT22	1μF	±20%, ±10%,±5%,	250V dc 160V ac	9V/μs	>2.5 G	1%	+400 ppm / °C	-55 to +100 °C	⌀ 0.8, 22.5, 19.5, 10, 26.25

Axial lead - metallized polyester capacitors

MKT

MANFR	TYPE No	C VALUE	STD TOL	MAX V	MAX dV/dt	INSUL R	Tan δ	TEMP CO (0-80°C)	TEMP RANGE	MAX DIMENSIONS
ARCO-TRONICS	MKT1.5	47nF	±20%,	250V dc 160V ac	40V/μs	>30 G	1.2% max at 10kHz	+400 ppm / °C	-55 to +100 °C	5.0, ⌀ .65, 11
ISKRA	KEU1012	47nF	±20%	250V dc	10V/μs	-	1.0% at 1kHz	-	-40 to +85 °C	5.5, ⌀ 0.6, 14
ROEDER-STEIN	MKT1813	47nF	±10%,±5%	250V dc	22V/μs	>30 G	0.8% at 1kHz	+375 ppm / °C	-55 to +100 °C	6.0, ⌀ 0.7, 14
THOMSON	S9 13	47nF	±10%	250V dc 100V ac	-	>7.5 G	0.8% at 1kHz	+400 ppm / °C	-55 to +100 °C	6.25, ⌀ 0.6, 14.5

Radial lead - metallized polypropylene capacitors

MKP

MANFR	TYPE No	C VALUE	STD TOL	MAX V	MAX dV/dt	INSUL R	Tan δ	TEMP CO (0-80°C)	TEMP RANGE	MAX DIMENSIONS
EVOX	PMR22.5	1μF	±20%, ±10%,±5% 2.5%, 2%	250V dc 160V ac	60V/μs	>30G	0.1% max at 10kHz	-300 ppm / °C	-55 to +105 °C	⌀ 0.8, 22.5, 16.5, 7.0, 26.5
PHILIPS	378	1μF	±5%	250V dc 160V ac	60V/μs	>100G	0.2% at 10kHz	-275 ppm / °C	-55 to +85 °C	⌀ 0.8, 27.5, 21, 11, 31
ROEDER-STEIN	MKP1840	1μF	±20%, ±10%,±5%,	250V dc 160V ac	75V/μs	>30G	0.06% at 10kHz	-300 ppm / °C	-55 to +100 °C	⌀ 0.8, 22.5, 18.5, 10.5, 26.5
SIEMENS	B32650	1μF	±10%,±5%,	250V dc 160V ac	25V/μs	>10G	0.08% at 10kHz	-300 ppm / °C	-55 to +85 °C	⌀ 0.8, 22.5, 16.5, 10.5, 26.5
THOMSON	BB22	1μF	±5%	250V dc 160V ac	195V/μs	>6G	0.1% at 1kHz	-300 ppm / °C	up to +100 °C	⌀ 0.8, 22.5, 21.5, 12.5, 26.25
WIMA	MKP10	1μF	±20%, ±10%,±5%,	250V dc 180V ac	65V/μs	>30G	0.06% at 10kHz	-300 ppm / °C	-55 to +100 °C	⌀ 0.8, 22.5, 21, 11, 26.5

3 Inductors

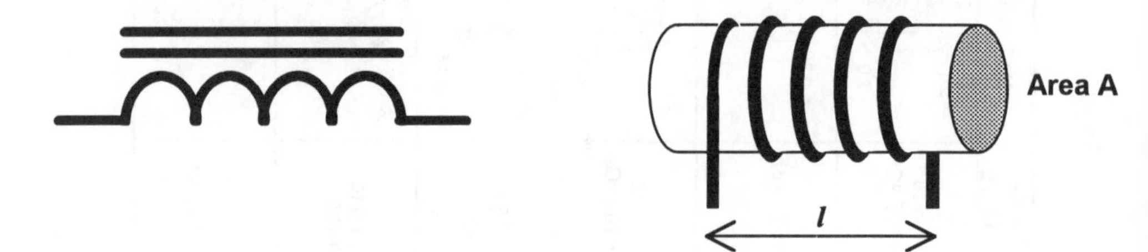

Definition

Inductance is the electrical property which exists due to the presence of a magnetic field around a conductor which is carrying current. Inductance may be thought of as electrical momentum. When there is appreciable inductance in a circuit it makes it difficult to build up electron flow or current quickly, but once the current is established, inductance makes it difficult to stop quickly. A single straight piece of wire has inductance although this is only in the order of 10nH per foot. Inductance may be increased by winding wire into a coil and may be increased substantially by winding the coil around a magnetic material such as iron. Inductance is increased in proportion to the number of turns squared.

Inductance of a long coil

$$L = \frac{\mu N^2 A}{l}$$

L = inductance in henrys
A = cross-sectional area of the coil in square metres
l = length of the coil in metres
μ = magnetic permeability of the coil's core (= $4\pi \times 10^{-7}$ for air)

This formula is only approximate because it assumes that all the magnetic flux lines link all the turns of the coil. In practice some of the flux lines fail to link some of the turns near the ends of the coil.

If the length of the coil is 10x its diameter, then the true inductance is 4% less than that given by the formula. If the length is 4x the diameter, the true inductance is 10% less than the formula.

Energy stored in an inductor

$$\text{Energy} = \frac{1}{2} L i^2$$

An inductor may be thought of as "storing" current (charge which is moving) compared to a capacitor which stores stationary charge. The energy is actually stored in the magnetic field and is proportional to the current squared.

Current build-up in an inductor

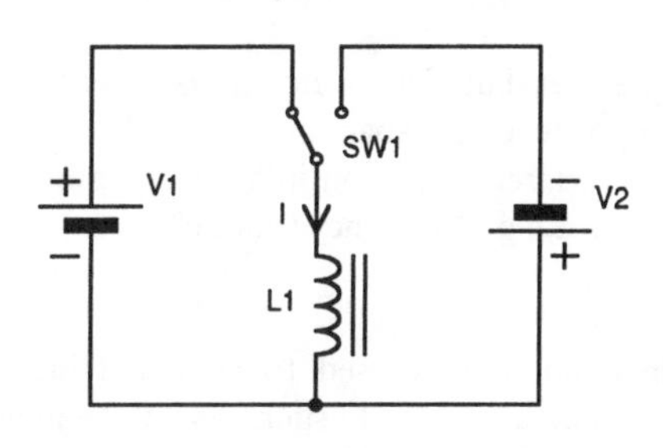

When the switch SW1 is in the position shown, V1 is connected to the inductor L1. Current builds up linearly according to the equation:

$$di/dt = V1/L1$$

After a while the current has built up and the switch is thrown over to the other position which connects L1 to the battery V2. The current cannot stop and continues to flow into V2. The battery V2 is forcing the current to stop and the current starts to linearly reduce according to the equation:

$$di/dt = V2 / L1$$

The current does not suddenly change its value, but starts to ramp down in magnitude at a rate of change dictated by the new battery voltage it is connected to. It should be mentioned that in practice, the switch should instantly change from one position to the other without a break.

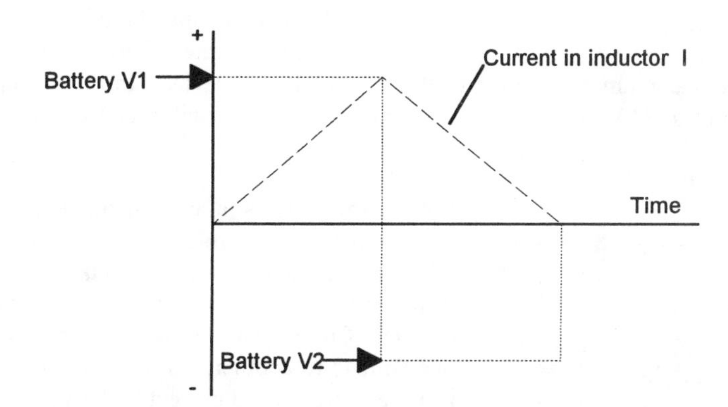

Inductor in an ac circuit

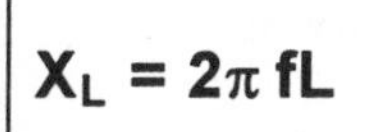

If the voltage across the inductor is changing in a sinusoidal fashion at a frequency f, then the current that flows also has a sinusoidal waveform. The amplitude of the sinusoidal current is proportional to the amplitude of the sinusoidal voltage across the inductor. The current $I = V/X_L$ where X_L is the reactance of the inductor in ohms.

Rules of thumb

- Inductance may be thought of as storing current (electrons which are moving).
- The current in an inductor cannot change instantly.
- The voltage across an inductor can change instantly and will do so to try and keep the current through it constant.
- An inductor can receive energy from a circuit which has a low dc voltage supply and can then release the energy at a much higher voltage. It cannot release the energy at a higher current than the final value the current was when it received the energy.
- In an ac circuit, the current lags the voltage by 90 degrees.
- The reactance of an inductor is $2 \pi f L$, where f is the frequency of the supply.

Uses of an inductor

AC blocking

A pure inductor has a high impedance at high frequencies but allows dc to pass without attenuation, consequently it may be considered the opposite of a capacitor. An inductor is often used to block the conduction of high frequency interference into an item of equipment. An inductor used in this fashion is sometimes referred to as a high frequency "choke".

di/dt limiting

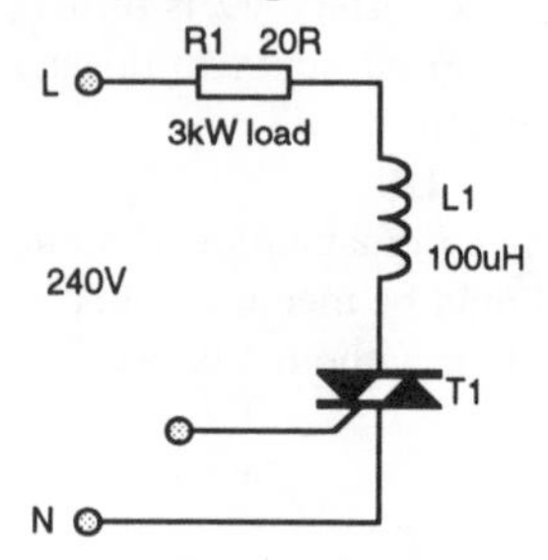

When a semiconductor is used to switch a high-power, low-inductance load such as a heating element, there may be a very high rate of increase of current when the device is turned on. High power semiconductors require a finite time to fully turn on. If the current through the device increases too quickly, it may "crowd" into one area of the silicon chip and cause overheating and destruction of the device. An inductor may be used in series with the device to limit the rate of rise of the current (di/dt) to less than the maximum di/dt rating of the semiconductor device. In the example opposite, L1 limits the di/dt to $240 \times 1.414/10^4 = 3.4$ amps per microsecond (worst-case).

Power conversion

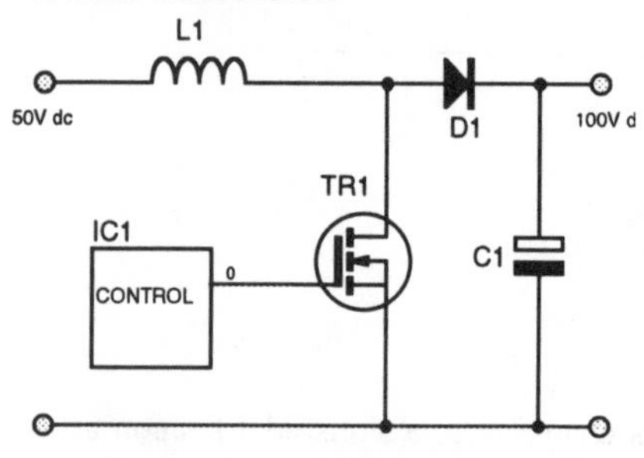

When current flows in an inductor, energy is stored in the magnetic field around it. This energy may be released at a higher or lower voltage than the voltage which was used in the process of building up the current. Consequently power may be converted from one voltage to another. In the example opposite, TR1 is turned hard-on and current builds up in L1 at the rate of 50V/L1. When TR1 is turned off, the current in L1 continues to flow through D1 into C1. By choosing the values of L1 and C1 and arranging for the appropriate on and off times of TR1, the output voltage may be boosted to 100V dc.

Resonate with a capacitor

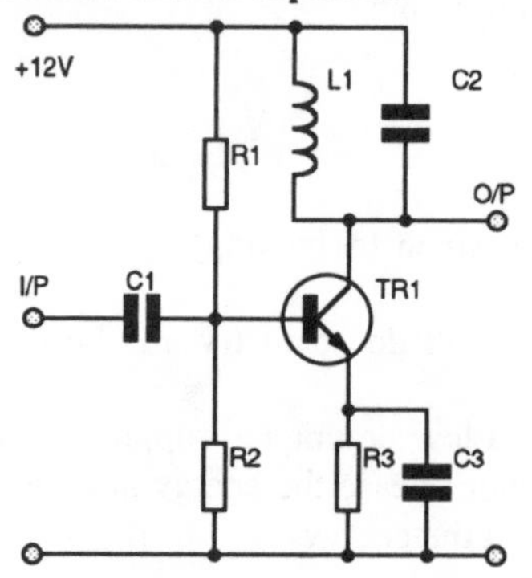

The circuit opposite is a tuned amplifier. It has the highest gain at one particular frequency; the resonant frequency of L1 and C2. At this frequency, the reactance of the inductor L1 ($2\pi fL1$) exactly equals the reactance of the capacitor C2 ($1/2\pi fC2$). At the resonant frequency, the combination of L1 and C2 presents a very high ac impedance to TR1's collector. The ac output voltage is therefore at a maximum.

Transformers

A transformer converts electrical energy at one ac voltage to another (usually different) ac voltage at the same frequency. The winding which is energized is called the primary and the winding or windings which receive energy from the primary are called secondaries. A transformer may be specifically designed to transform power, voltage or current but the same basic concepts apply.

The perfect transformer

The perfect transformer is 100% efficient. It transforms an ac voltage applied to the primary to an ac voltage at the secondary in exact proportion to the turns ratio between the primary and the secondary. The output voltage is not affected by the current drawn.

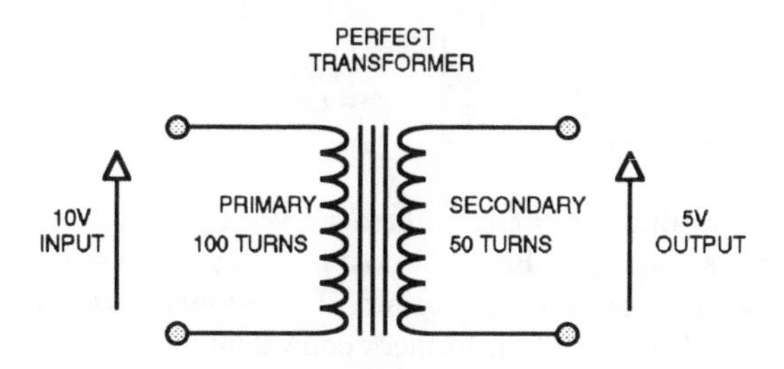

$$\text{Output voltage} = \text{Input voltage} \times \frac{\text{Secondary turns}}{\text{Primary turns}}$$

In practice, a transformer is not perfect due to losses in the copper wire and the iron core. Its output voltage is affected by the current drawn due to voltage drops in the copper wire and voltage drops in the leakage reactances.

Transformer equation

$$\text{Induced voltage} = 4.44\ N f \Phi$$

N = number of turns
f = frequency
Φ = magnetic flux

Rules of thumb

- Flux is determined by: a) the applied voltage,
  b) the frequency,
  c) the number of primary turns.
- The flux is not affected by load current.
- The magnetizing current is determined by the core construction; it must adjust so as to produce a flux to give an induced voltage equal to the applied voltage.
- A transformer may be used on a higher frequency.
- It may not be used on a lower frequency.
- A transformer on no-load is an inductance (magnetizing) in parallel with a high value resistance.(core losses).
- A transformer is most efficient when core losses = copper losses.

Transformer equivalent circuit

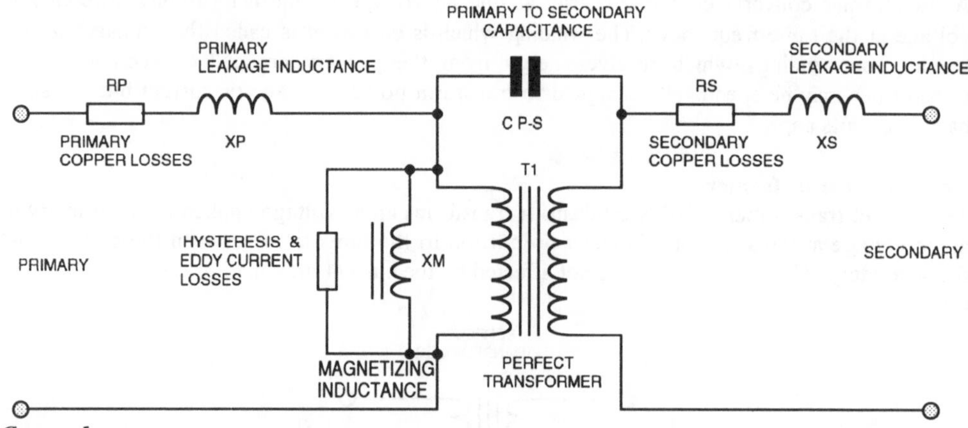

Copper losses

When current is drawn from the secondary, there are voltage drops generated in the resistance of the copper wire of the windings. Energy is lost in heat and the secondary voltage falls from the value it would be in a perfect transformer. The copper losses in the primary may be represented by a resistor, Rp, in the transformer's equivalent circuit.

Core losses

Energy is lost in the magnetic core of a transformer due to eddy currents and hysteresis. The losses may be represented by a resistor, Rc, in the equivalent circuit.

Magnetizing inductance

When no current is being drawn from the secondary, the primary appears as an inductance which has a reactance (Xm) at the frequency of the supply to which it is connected. A current is drawn from the supply due to Xm and is called the magnetizing current.

Leakage inductances

Some of the magnetic flux produced by the primary current does not link with the secondary winding. It is termed leakage flux and may be equated to an inductance called the primary leakage inductance. At the supply frequency it will have a reactance, Xp, and will cause a voltage drop in the secondary voltage when current is drawn. The secondary also has a leakage inductance which will give a reactance, Xs. This will also cause a voltage drop when current is drawn. The leakage effects are mainly in air and so the leakage inductances will not saturate as the current in the windings increases.

Inter-winding capacitance

There will be stray capacitance between the primary and secondary windings due to the fact that they are close to each other. A current will flow through this capacitance although generally it is a common mode effect which will cause leakage currents to earth in mains transformer systems. Modern mains transformers have their primary and secondary windings on separate bobbins which increases leakage inductances but reduces inter-winding capacitance.

Difference between a voltage transformer and a current transformer

Although current transformers are specially designed for the purpose, in theory, there is no difference between a current transformer and a voltage transformer. It is the way they are connected which dictates their type.

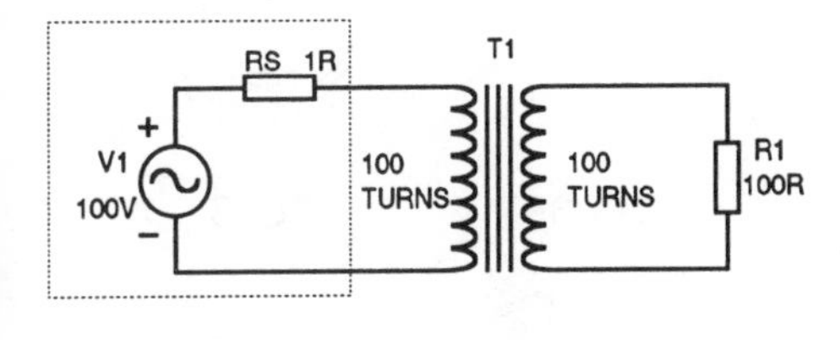

The circuit opposite employs an isolating transformer T1 with 1:1 turns ratio. The impedance of the voltage source V1 is 1 ohm. The 100 ohm load resistor R1, causes a current to flow of approximately 1A. 1A also flows in the primary. It is the transformer and its load which are dictating the current which flows from the source. T1 is connected as a voltage transformer as it is presenting the source voltage to the resistor R1.

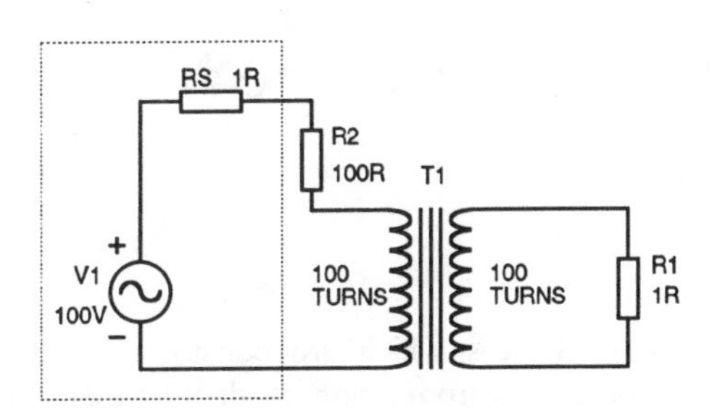

The same transformer is now connected with its primary in series with a 100 ohm resistor. The secondary has a 1 ohm load. It is now the 100 ohm resistor R2 which is dictating the current flowing from the voltage source as the impedance presented by the transformer is much lower than R2. T1 is now connected as a current transformer as it is transforming the current set by R2 across to the resistor R1.

A current transformer can be removed from a system and replaced with a short circuit and the remaining system will be unaffected. This cannot be done with a voltage transformer.

In practice, current transformers are designed with a low number of turns on the primary (often 1/2 turn) and a much larger number of turns on the secondary. They are used to enable an ammeter of say 1A full scale to measure many hundreds of amps in a circuit where the current is dictated by other devices in series with the transformer and not the transformer itself.

4 Semiconductors

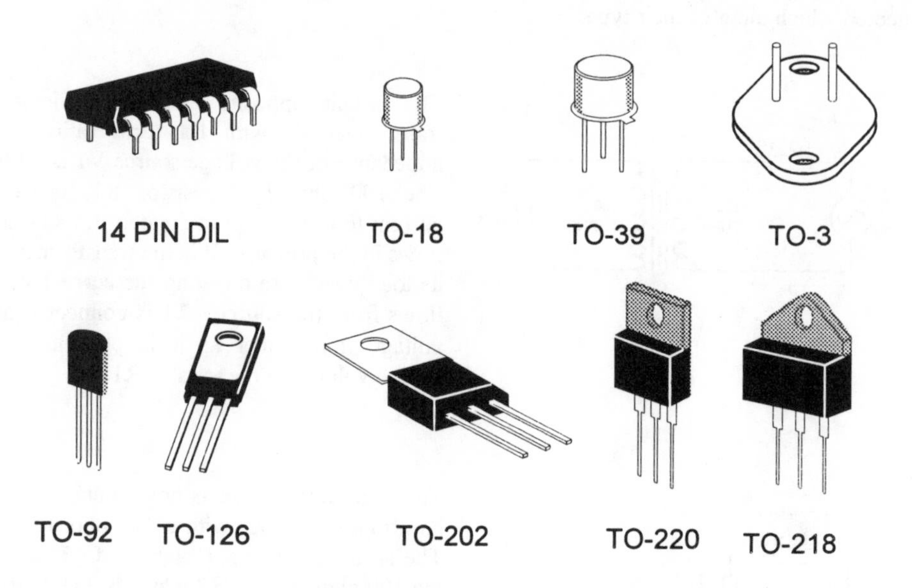

Semiconductors is the collective term for electronic devices which are constructed from semiconductor material. Semiconductor material is processed from a non-conducting material such as silicon and after processing it becomes conductive with a resistivity in between that of a metallic conductor (such as copper) and an insulator such as ceramic. Materials of different types may be combined to form devices such as transistors, diodes, thyristors and integrated circuits.

This section contains basic data and concepts related to semiconductor devices which are in common use. The cross-reference tables at the end of each subsection include devices which have evolved as "industry standard" parts. They have been selected as being the most common devices which appear in electronic component distributors' catalogues.

In general, semiconductor devices are less linear than passive components such as resistors and capacitors i.e. the current in a semiconductor device does not linearly follow the voltage applied across it.

Semiconductor devices are electrically less rugged than passive components. Semiconductors may be destroyed by too much heat during the soldering process and may be damaged by static electricity during handling.

Semiconductor prefixes

A common problem experienced by engineers is to identify the manufacturer of a semiconductor device from its type number only. The following table gives an alphabetical listing of letter combinations and the manufacturers of the devices having type numbers commencing with those letters.

1.5KE	SGS
A	AMD
ACF	GENERAL INSTRUMENT
ACP	ITT
ACR	MEDL
ACVP	ITT
AD	ANALOG DEVICES
ADC	ANALOG DEVICES, HARRIS SEMICONDUCTOR, GENERAL INSTRUMENT, ITT
AH	NATIONAL SEMICONDUCTOR
AM	ADVANCED MICRO DEVICES, AMD
AMU	ITT
AN	PANASONIC (MATSUSHITA)
APU	ITT
AR	GENERAL INSTRUMENT
ARS	GENERAL INSTRUMENT
ASC	ITT
AW	GENERAL INSTRUMENT
AY	GENERAL INSTRUMENT
B	FAGOR, GENERAL INSTRUMENT
BA	FAGOR
BAT	ITT, PHILIPS COMPONENTS
BAV	ITT, PHILIPS
BAW	ITT
BAX	ITT
BB	ITT
BY	FAGOR
BYM	GENERAL INSTRUMENT
BYV	GENERAL INSTRUMENT
BYW	GENERAL INSTRUMENT
BDM	ANALOG DEVICES
BSS	SIEMENS
BT	PHILIPS COMPONENTS
BTA	THOMSON
BTB	THOMSON
BTW	THOMSON
BUD	THOMSON
BUF	THOMSON
BUH	THOMSON
BUL	THOMSON, MOTOROLA
BUR	THOMSON
BUT	THOMSON
BUV	THOMSON
BUW	THOMSON
BUX	THOMSON
BUZ	SIEMENS, HARRIS SEMICONDUCTOR, THOMSON
BY	THOMSON
BYT	THOMSON
BYV	THOMSON
BYW	THOMSON
BYX	THOMSON
BZV	THOMSON

BZW	THOMSON
BZX	THOMSON
BZY	THOMSON
C	GENERAL INSTRUMENT, AMD
CA	HARRIS SEMICONDUCTOR, FAIRCHILD, PHILIPS COMPONENTS, INTERNATIONAL RECTIFIER
CAP	ITT
CB	THOMSON, INTERNATIONAL RECTIFIER
CBC	INTERNATIONAL RECTIFIER
CC	GENERAL INSTRUMENT
CCD	ITT
CCU	ITT
CD	NATIONAL SEMICONDUCTOR, HARRIS SEMICONDUCTOR,
CDD	FAIRCHILD
CDG	TELEDYNE
CDP	HARRIS SEMICONDUCTOR
CF	GENERAL INSTRUMENT
CG	STA, GENERAL INSTRUMENT
CK	GENERAL INSTRUMENT
CL	GENERAL INSTRUMENT
CLK	ITT
CMP	PRECISION MONOLITHICS INCORPORATED
CN	THOMSON
CP	ITT,GENERAL INSTRUMENT
CPA	INTERNATIONAL RECTIFIER
CPAT	INTERNATIONAL RECTIFIER
CPT	INTERNATIONAL RECTIFIER
CPU	INTERNATIONAL RECTIFIER
CPV	INTERNATIONAL RECTIFIER
CPY	INTERNATIONAL RECTIFIER
COM	STA
CR	SILICONIX
CS	INTERNATIONAL RECTIFIER, NIPPON ELECTRIC CO
CSP	ITT
CT	PLESSEY, INTERNATIONAL RECTIFIER
CTC	INTERNATIONAL RECTIFIER
CVPU	ITT
CX	SONY
CXA	SONY
CXB	SONY
CXD	SONY
CXK	SONY
CXL	SONY
D	SILICONIX, HARRIS SEMICONDUCTOR, THOMSON
DAC	GENERAL INSTRUMENT, DAT, ANALOG DEVICES, PRECISION MONOLITHICS INCORPORATED, NATIONAL SEMICONDUCTOR
DAS	ANALOG DEVICES
DB	THOMSON
DC	THOMSON
DF	GENERAL INSTRUMENT, INTERNATIONAL RECTIFIER
DG	SILICONIX, GENERAL INSTRUMENT, HARRIS SEMICONDUCTOR
DH	NATIONAL SEMICONDUCTOR
DM	NATIONAL SEMICONDUCTOR
DMA	ITT
DN	NATIONAL SEMICONDUCTOR, PANASONIC, NIPPON ELECTRIC CO
DNO	NIPPON ELECTRIC CO
DP	NATIONAL SEMICONDUCTOR, INTERNATIONAL RECTIFIER

DPU	ITT
DS	NATIONAL SEMICONDUCTOR
DSC	ANALOG DEVICES
DTI	ITT
DTM	ANALOG DEVICES
DTV	THOMSON
DVR	PRECISION MONOLITHICS INCORPORATED
E	THOMSON
EDF	GENERAL INSTRUMENT
EF	THOMSON
EFG	THOMSON
EFR	THOMSON
EGL	GENERAL INSTRUMENT
EGP	GENERAL INSTRUMENT
EL	ELAN
EM	ITT
EMUS	THOMSON
EP	PANASONIC (MATSUSHITA)
ER	GENERAL INSTRUMENT
ESM	THOMSON
ETC	THOMSON
ETL	THOMSON
F	FAIRCHILD
FA	NEC
FAOH	NEC
FB	FAGOR
FBP	FAGOR
FBU	FAGOR
FC	NEC
FCD	FAIRCHILD
FCM	FAIRCHILD
FCS	FAIRCHILD
FE	GENERAL INSTRUMENT
FEP	GENERAL INSTRUMENT
FES	GENERAL INSTRUMENT
FIP	NEC
FLC	FAIRCHILD
FLV	FAIRCHILD
FM	NEC
FNA	FAIRCHILD
FND	FAIRCHILD
FPA	FAIRCHILD
FPE	FAIRCHILD
FPT	FAIRCHILD
FSOH	NEC
FZOH	NEC
G	SILICONIX, GENERAL INSTRUMENT
GAL	THOMSON, LATTICE SEMICONDUCTOR
GBPC	GENERAL INSTRUMENT
GBU	GENERAL INSTRUMENT
GE	HARRIS SEMICONDUCTOT
GEM	HARRIS SEMICONDUCTOR CO
GES	HARRIS SEMICONDUCTOR
GI	GENERAL INSTRUMENT
GL	GENERAL INSTRUMENT
GLL	GENERAL INSTRUMENT

GP	GENERAL INSTRUMENT
GPP	GENERAL INSTRUMENT
GS	THOMSON
GSC	THOMSON
GSD	THOMSON
GT	SESCOSEM
H	THOMSON,FAIRCHILD, HARRIS SEMICONDUCTOR, HITACHI
HA	HITACHI, HARRIS SEMICONDUCTOR
HAL	ITT
HB	HITACHI
HD	HITACHI
HE	INTERNATIONAL RECTIFIER
HEF	PHILIPS COMPONENTS
HBF	THOMSON
HC	HARRIS SEMICONDUCTOR,MHS
HCC	HARRIS SEMICONDUCTOR
HCF	HARRIS SEMICONDUCTOR, THOMSON
HCPL	HEWL
HCTL	HEWL
HD	HARRIS SEMICONDUCTOR, HITACHI
HDSP	HEWL
HEDS	HEWL
HEMT	HEWL
HFA	HARRIS SEMICONDUCTOR, INTERNATIONAL RECTIFIER
HFBR	HEWL
HI	HARRIS,SEMICONDUCTOR
HLCP	HEWL
HLMP	HEWL
HM	HARRIS SEMICONDUCTOR, MHS, HITACHI
HN	HITACHI
HSMP	HEWL
HSMS	HEWL
HSMX	HEWL
HSMY	HEWL
HSSR	HEWL
HTA	THOMSON
HTIL	HEWL
HV	HARRIS SEMICONDUCTOR
HXTR	HEWL
IC	NEC
ICL	HARRIS SEMICONDUCTOR
ICM	HARRIS SEMICONDUCTOR
ID	HESMI
IGT	HARRIS SEMICONDUCTOR
IH	HARRIS SEMICONDUCTOR
IL	LITRONIX
IM	HARRIS SEMICONDUCTOR. INMOS
IMS	THOMSON
IMST	INMOS
IP	HARRIS SEMICONDUCTOR
IR	INTERNATIONAL RECTIFIER
IRC	INTERNATIONAL RECTIFIER
IRCC	INTERNATIONAL RECTIFIER
IRCP	INTERNATIONAL RECTIFIER
IRCZ	INTERNATIONAL RECTIFIER
IRD	INTERNATIONAL RECTIFIER
IRF	INTERNATIONAL RECTIFIER, HARRIS SEMICONDUCTOR, THOMSON

IRF	INTERNATIONAL RECTIFIER
IRG	INTERNATIONAL RECTIFIER
IRH	INTERNATIONAL RECTIFIER
IRK	INTERNATIONAL RECTIFIER
IRL	INTERNATIONAL RECTIFIER
IRT	ITT
IS	HARRIS SEMICONDUCTOR
ISB	THOMSON
ISP	ITT
IT	HARRIS SEMICONDUCTOR
ITA	THOMSON
ITT	ITT
J	HARRIS SEMICONDUCTOR, SILICONIX
JA	ITT
JAN	INTERNATIONAL RECTIFIER
JC	ITT
K	INTERNATIONAL RECTIFIER
KA	SAMSUNG
KB	GENERAL INSTRUMENT
KBPC	INTERNATIONAL RECTIFIER, MICRO ELECTRONICS
KR	STANDARD MICROSYSTEMS (US)
L	THOMSON,SILICONIX
LA	LAMBDA
LCP	THOMSON
LDP	THOMSON
LD	HARRIS SEMICONDUCTOR
LF	TEXAS INST, NATIONAL SEMICONDUCTOR, PHILIPS COMPONENTS, MOTOROLA, THOMSON
LH	NATIONAL SEMICONDUCTOR, PHILIPS COMPONENTS, SILICONIX
LL	ITT
LM	NATIONAL SEMICONDUCTOR, TEXAS INSTRUMENTS, FAIRCHILD, PHILIPS COMPONENTS, SILICONIX, LINEAR TECH, THOMSON, HARRIS
LP	HARRIS SEMICONDUCTOR CO, GENERAL INSTRUMENT, NATIONAL SEMICONDUCTOR
LS	THOMSON
LT	LINEAR TECH
LTC	LINEAR TECH
LTK	LINEAR TECH
LZ	ITT
M	THOMSON, MITSUBISHI
MAB	PHILIPS COMPONENTS
MAC	MOTOROLA
MAF	PHILIPS COMPONENTS
MAX	MAXIM
MB	RECTRON
MBD	MOTOROLA
MBR	GENERAL INSTRUMENT, INTERNATIONAL RECTIFIER, MOTOROLA
MC	MOTOROLA, HARRIS SEMICONDUCTOR CO, GENERAL INSTRUMENT, THOMSON, NEC
MCM	MOTOROLA
MCR	MOTOROLA
MCT	MOTOROLA
MCU	ITT
MD	MOTOROLA
MDA	ITT, MOTOROLA
MDC	MOTOROLA

MDS	THOMSON
MEM	HARRIS SEMICONDUCTOR CO, GENERAL INSTRUMENT
MFE	MOTOROLA
MFO	MOTOROLA
MH	NATIONAL SEMICONDUCTOR
MJ	MOTOROLA, THOMSON, HARRIS SEMICONDUCTOR
MJD	MOTOROLA
MJE	MOTOROLA, HARRIS SEMICONDUCTOR, THOMSON
MJF	MOTOROLA
MJH	MOTOROLA
MIC	ITT
MK	MOSTEK, THOMSON, MOTOROLA
ML	PLESSEY, THOMSON
MLED	MOTOROLA
MLM	MOTOROLA
MM	NATIONAL SEMICONDUCTOR, LINEAR TECH
MM	MOTOROLA
MN	PANASONIC
MOC	MOTOROLA
MOD	SILICONIX
MP	TOSHIBA
MPF	MOTOROLA
MPM	MOTOROLA
MPN	MOTOROLA
MPQ	MOTOROLA
MPG	GENERAL INSTRUMENT
MPQ	THOMSON
MPS	MOTOROLA
MPSA	MOTOROLA
MPSH	MOTOROLA
MPSL	MOTOROLS
MPSU	MOTOROLA
MPSW	MOTOROLA
MPTE	MOTOROLA
MPX	ANALOG DEVICES, MOTOROLA
MPF	NATIONAL SEMICONDUCTOR
MPY	BURR BROWN
MR	MOTOROLA
MRD	MOTOROLA
MRF	MOTOROLA
MSD	MOTOROLA
MSE	ITT
MSM	OKI
MSP	ITT
MSS	THOMSON
MTA	MOTOROLA
MTD	MOTOROLA
MTE	MOTOROLA
MTH	MOTOROLA, THOMSON
MTM	MOTOROLA
MTP	MOTOROLA, THOMSON
MR	FAGOR
MU	MOTOROLA
MUR	HARRIS SEMICONDUCTOR, MOTOROLA
MUX	HARRIS SEMICONDUCTOR CO, GENERAL INSTRUMENT
MV	MOTOROLA
MVAM	MOTOROLA
MZD	MOTOROLA

N	PHILIPS COMPONENTS
NA	HARRIS SEMICONDUCTOR
NE	PHILIPS COMPONENTS, THOMSON
NOM	PLESSEY
NP	GENERAL INSTRUMENT
NPF	GENERAL INSTRUMENT
NR	HARRIS SEMICONDUCTOR
NS	GENERAL INSTRUMENT
NSF	GENERAL INSTRUMENT
NVM	ITT
OP	PLESSEY, PRECISION MONOLITHICS INCORPORATED
P	INTERNATIONAL RECTIFIER, AMD, GENERAL INSTRUMENT, THOMSON, HARRIS, MOTOROLA
PA	PHILIPS COMPONENTS
PBL	THOMSON
PCA	PHILIPS COMPONENTS
PCB	PHILIPS COMPONENTS
PCD	PHILIPS COMPONENTS
PCF	PHILIPS COMPONENTS
PFR	THOMSON
PIC	ITT, HARRIS SEMICONDUCTOR CO, GENERAL INSTRUMENT, UNITRODE
PIP	ITT
PL	THOMSON
PLQ	THOMSON
PM	HARRIS SEMICONDUCTOR CO, GENERAL INSTRUMENT, PRECISION MONOLITHICS INCORPORATED
PNA	PHILIPS COMPONENTS
PP	HARRIS SEMICONDUCTOR CO, GENERAL INSTRUMENT
PS	NEC
PSP	ITT
PVA	INTERNATIONAL RECTIFIER
PVD	INTERNATIONAL RECTIFIER
PVI	INTERNATIONAL RECTIFIER
PVPU	ITT
PVR	INTERNATIONAL RECTIFIER
QFT	TEL
R	INTERNATIONAL RECTIFIER
RA	HARRIS SEMICONDUCTOR CO, GENERAL INSTRUMENT, RAYTHEON
RAY	RAYTHEON
RC	RAYTHEON
RCA	HARRIS SEMICONDUCTOR
RCR	ITT
RCT	ITT
REF	PRECISION MONOLITHICS INCORPORATED
REG	BURR BROWN
RF	HARRIS SEMICONDUCTOR
RG	GENERAL INSTRUMENT
RGL	GENERAL INSTRUMENT
RGP	GENERAL INSTRUMENT
RL	RECTRON
RM	HARRIS SEMICONDUCTOR CO, GENERAL INSTRUMENT, RAYTHEON
RMP	GENERAL INSTRUMENT
RO	ITT, HARRIS SEMICONDUCTOR CO, GENERAL INSTRUMENT
RRF	HARRIS SEMICONDUCTOR
RS	GENERAL INSTRUMENT
RTI	ANALOG DEVICES

RUR	HARRIS SEMICONDUCTOR
S	SIEMENS, TAG SEMICONDUCTORS, SEIKO INSTRUMENTS, INTERNATIONAL RECTIFIER
SA	PHILIPS COMPONENTS, THOMSON
SAA	PHILIPS COMPONENTS, ITT, HARRIS SEMICONDUCTOR CO, GENERAL INSTRUMENT, THOMSON
SAB	PHILIPS COMPONENTS
SAC	ANALOG DEVICES
SAD	ITT
SAF	ITT
SAK	ITT
SAJ	ITT,THOMSON
SAH	ITT
SAS	SIEMENS
SB	GENERAL INSTRUMENT
SBA	HARRIS SEMICONDUCTOR CO, GENERAL INSTRUMENT
SBL	GENERAL INSTRUMENT
SBLF	GENERAL INSTRUMENT
SBP	TEXAS INSTRUMENTS
SC	INTERNATIONAL RECTIFIER
SD	PHILIPS COMPONENTS, TELEDYNE, INTERNATIONAL RECTIFIER, ITT
SDA	THOMSON
SDC	ANALOG DEVICES
SE	PHILIPS COMPONENTS, THOMSON
SFC	SESCOSEM
SG	SILICONIX, HARRIS SEMICONDUCTOR, SILICON GENERAL, PHILIPS, SPRAGUE, THOMSON
SGL	GENERAL INSTRUMENT
SGS	THOMSON
SGT	HARRIS SEMICONDUCTOR
SH	FAIRCHILD
SHA	ANALOG DEVICES
SI	SILICONIX
SKB	SEMIKRON
SKD	SEMIKRON
SKKD	SEMIKRON
SKKH	SEMIKRON
SKKT	SEMIKRON
SL	HARRIS SEMICONDUCTOR CO, GENERAL INSTRUMENT, PLESSEY
SM	SILICON GENERAL, THOMSON, SEMELAB
SML	SEMELAB
SMM	SILICONIX
SMP	PRECISION MONOLITHICS INCORPORATED, SILICONIX
SMV	SILICONIX
SMW	SILICONIX
SMX	ANALOG DEVICES
SN	TEXAS INSTRUMENTS
SO	THOMSON
SP	PLESSEY, INTERNATIONAL RECTIFIER, HARRIS SEMICONDUCTOR
SPBM	SILICONIX
SPMF	SILICONIX
SPU	ITT
SSC-	INTERNATIONAL RECTIFIER
SSM	PRECISION MONOLITHICS INCORPORATED
SR	STANDARD MICROSYSTEMS CORP
SRP	GENERAL INSTRUMENT
SRX	ANALOG DEVICES
SS	HARRIS SEMICONDUCTOR CO, GENERAL INSTRUMENT

SSC	ANALOG DEVICES
SSS	PRECISION MONOLITHICS INCORPORATED
ST	THOMSON, INTERNATIONAL RECTIFIER
STF	THOMSON
STGH	THOMSON
STGP	THOMSON
STH	THOMSON
STHV	THOMSON
STK	THOMSON
STKM	THOMSON
STLT	THOMSON
STP	THOMSON
STPR	THOMSON
STPS	THOMSON
STV	THOMSON
STVH	THOMSON
SU	PHILIPS COMPONENTS
SW	PLESSEY
T	HARRIS SEMICONDUCTOR CO, GENERAL INSTRUMENT,TRANSITRON,THOMSON, TAG, INTERNATIONAL RECTIFIER
TA	TOSHIBA
TAA	THOMSON,PHILIPS COMPONENTS,ITT,SIEMENS,FAIRCHILD
TBA	THOMSON,PHILIPS COMPONENTS,ITT,SIEMENS,FAIRCHILD, TELFUNKEN
TBB	SIEMENS
TBC	SIEMENS
TBE	SIEMENS
TCA	PHILIPS COMPONENTS,ITT,SIEMENS, TELEFUNKEN
TC	TOSHIBA, TELEDYNE
TD	TOSHIBA
TDA	PHILIPS COMPONENTS, ITT, SIEMENS, FAIRCHILD, THOMSON, TELEFUNKEN
TDB	SIEMENS, THOMSON
TDC	SIEMENS, THOMSON
TDD	PHILIPS COMPONENTS
TDE	THOMSON
TDF	THOMSON
TDP	THOMSON
TEA	PHILIPS COMPONENTS, TELEFUNKEN, THOMSON
TEB	THOMSON
TEF	THOMSON
TFF	TRANSITRON
TFIR	ITT
TFPO	ITT
TGAL	THOMSON
TGDV	THOMSON
TGF	THOMSON
TH	TOSHIBA
THBT	THOMSON
THDT	THOMSON
THM	TOSHIBA
TIP	TEXAS INSTRUMENTS, THOMSON, HARRIS SEMICONDUCTOR
TF	TRANSITRON
TG	TRANSITRON
TGL	GENERAL INSTRUMENT
TL	TEXAS INSTRUMENTS, THOMSON
TLC	TEXAS INSTRUMENTS, THOMSON
TLG	TOSHIBA
TLN	TOSHIBA
TLO	TOSHIBA
TLP	TOSHIBA

TLS	THOMSON
TLR	TOSHIBA
TLRA	TOSHIBA
TLRC	TOSHIBA
TLS	TOSHIBA
TLSG	TOSHIBA
TLUR	TOSHIBA
TLY	TOSHIBA
TM	TOSHIBA, THOMSON
TMM	THOMSON
TMP	TOSHIBA
TMS	TEXAS INSTRUMENTS
TNG	TRANSITRON
TOSH	TOSHIBA
TP	TELEDYNE
TPQ	SPRAGUE
TPS	TOSHIBA
TPU	ITT
TOA	TRANSITRON
TODV	THOMSON
TPA	THOMSON
TPB	THOMSON
TPDV	THOMSON
TPP	THOMSON
TRAL	THOMSON
TRC	TRANSITRON
TS	THOMSON
TSA	PHILIPS COMPONENTS
TSD	THOMSON
TSDC	ANALOG DEVICES
TSFJ	THOMSON
TSFK	THOMSON
TSG	THOMSON
TSGF	THOMSON
TSGS	THOMSON
TSL	ANALOG DEVICES
TVPO	ITT
TXDV	THOMSON
TYN	THOMSON
TYP	THOMSON
TYS	THOMSON
TZ	HARRIS SEMICONDUCTOR, GENERAL INSTRUMENT
U	FAIRCHILD, TELEFUNKEN
UA	FAIRCHILD, PHILIPS COMPONENTS, TEXAS INSTRUMENTS, NATIONAL SEMICONDUCTOR, THOMSON
UAA	ITT, SIEMENS, TELEFUNKEN, THOMSON
UAF	ITT, THOMSON
UC	UNITRODE, THOMSON, LINEAR TECH
UCN	SPRAGUE
UCS	SPRAGUE
UDN	PHILIPS COMPONENTS, SPRAGUE
UDS	SPRAGUE
UF	GENERAL INSTRUMENT
UHD	SPRAGUE
UHP	SPRAGUE
ULC	SPRAGUE
ULN	SPRAGUE, PHILIPS COMPONENTS, THOMSON
ULQ	THOMSON

ULS	SPRAGUE
UPB	NEC
UPC	STANDARD MICROSYSTEMS, NIPPON ELECTRIC COMPANY
UPD	NEC
V	HARRIS SEMICONDUCTOR
VAD	ITT
VB	THOMSON
VCU	ITT
VDU	ITT
VSP	ITT
VN	TELEDYNE, THOMSON
W	GENERAL INSTRUMENT
X	TAG SEMICONDUCTORS
XR	EXAR
Y	TEXAS INSTRUMENTS
TYF	TOSHIBA
YTFP	TOSHIBA
Z	THOMSON, INTERNATIONAL RECTIFIER
ZA	ZETEX, TEXAS INSTRUMENTS
ZAD	ZETEX, TEXAS INSTRUMENTS
ZD	INTERNATIONAL RECTIFIER
ZEL	ZETEX,TEXAS INSTRUMENTS
ZGL	GENERAL INSTRUMENT
ZGP	GENERAL INSTRUMENT
ZM	ITT
ZMM	ITT
ZMU	ITT
ZMY	ITT
ZN	ZETEX, FERRANTI
ZPD	ITT
ZPU	ITT
ZPY	ITT
ZR	ZETEX
ZS	FERRANTI, PLESSEY
ZTE	ITT
ZTX	ITT
ZY	ITT
ZZ	ITT
ZZY	ITT

Note:
Intersil and RCA are now Harris. Signetics is now Philips Components.
Ferranti transistors are now Zetex

Japanese transistor prefixes

2SA	PNP HIGH FREQUENCY
2SB	PNP LOW FREQUENCY
2SC	NPN HIGH FREQUENCY
2SD	NPN LOW FREQUENCY
2SH	UNIJUNCTION
2SJ	P CHANNEL FET
2SK	N CHANNEL FET

Occasionally the "2S" is omitted thus a PNP high frequency transistor may be marked "A1279" instead of "2SA1279".

Diodes

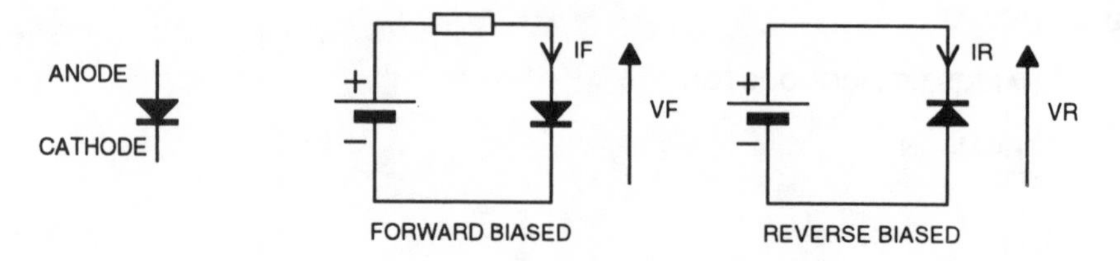

The diode is a two-terminal semiconductor device which conducts current readily when the anode is made positive with respect to the cathode (forward biased). A very small current flows when the cathode is made positive with respect to the anode (reverse biased).

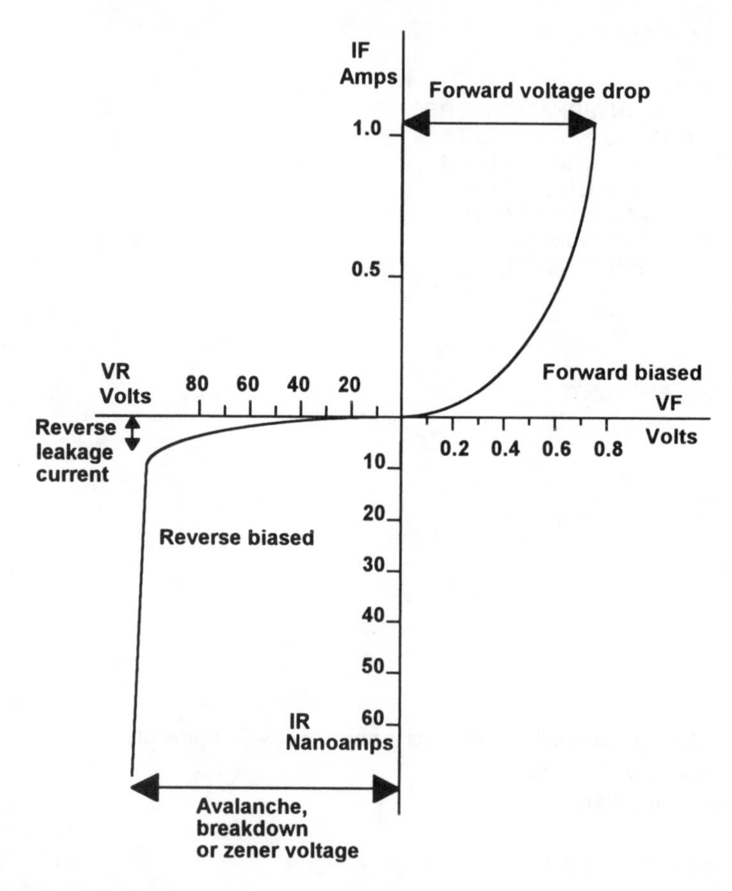

Rules of thumb

- A diode has a constant forward voltage drop = 0.6V.
- The reverse leakage current of a signal diode is approx 25nA at 25V.
- The forward drop reduces by 2.5mV for every 1 ^{o}C rise.
- The reverse leakage current doubles for every 10 ^{o}C rise.
- The impedance of a diode is approx = 25/I ohms (where I is the current in mA).

Uses of a diode

Rectification

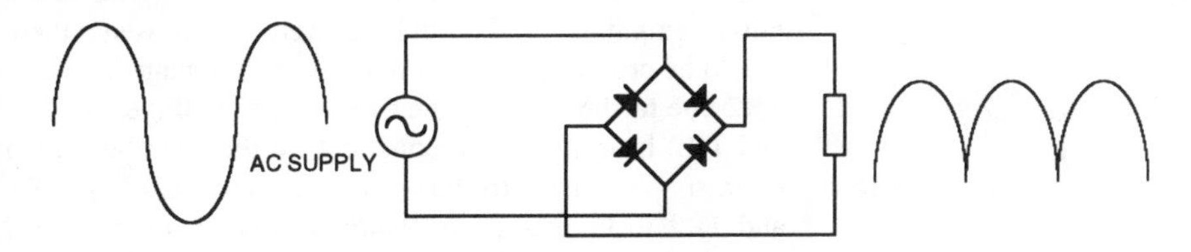

Auctioning

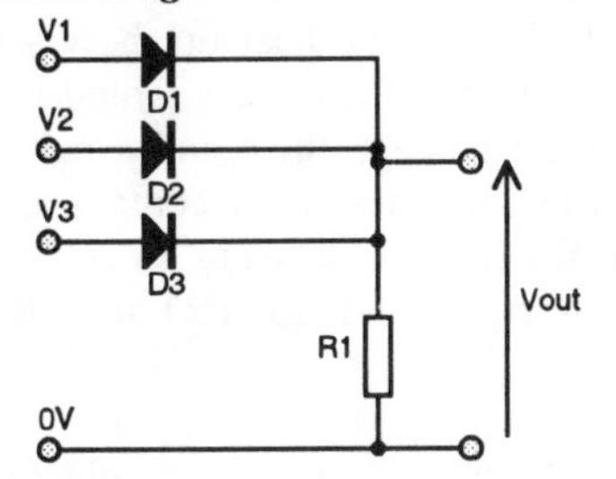

The output voltage, Vout, is always the highest positive voltage between V1, V2 or V3. When V1, V2 and V3 are digital signals, the circuit operates as an OR gate where Vout is only a logic 1 when V1, V2 or V3 is logic 1.

Clamping

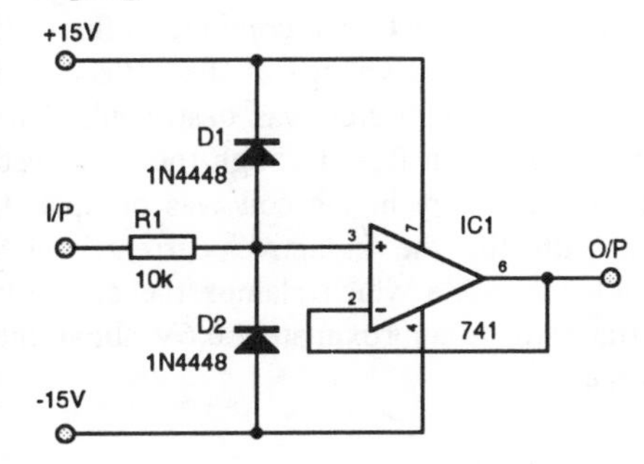

IC1 is connected as a voltage follower. The IC may be damaged if the input voltage exceeds the +15V and -15V supply rails. To prevent this happening, D1 and D2 clamp the input voltage to the +15V rail and the -15V rail respectively. R1 limits the current during the clamping process.

Creating a voltage pedestal

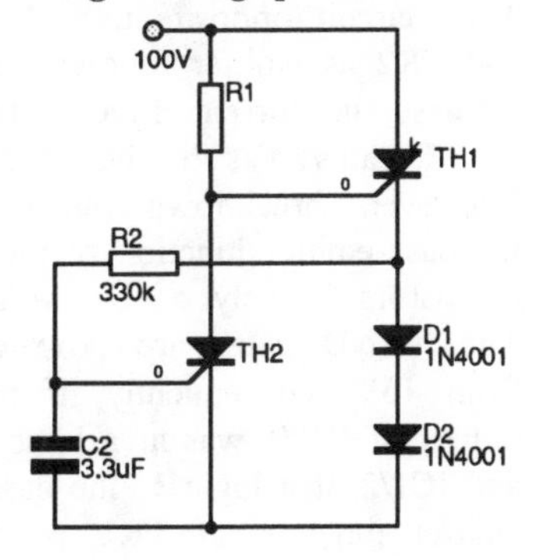

The forward drop of a diode is approximately 0.6V and may be considered constant for varying currents through it. The forward drop may be used as a crude 0.6V zener to create a voltage source or pedestal. In the circuit opposite, two diodes are used D1 and D2 to create a voltage for the time delay circuit R2 - C2. When the gate turn-off thyristor TH1 is triggered, approximately 1.2V appears across D1 and D2. C2 charges via R2 and when the gate threshold voltage of TH2 is reached, TH2 is triggered and turns TH1 off by GTO action.

Temperature compensation

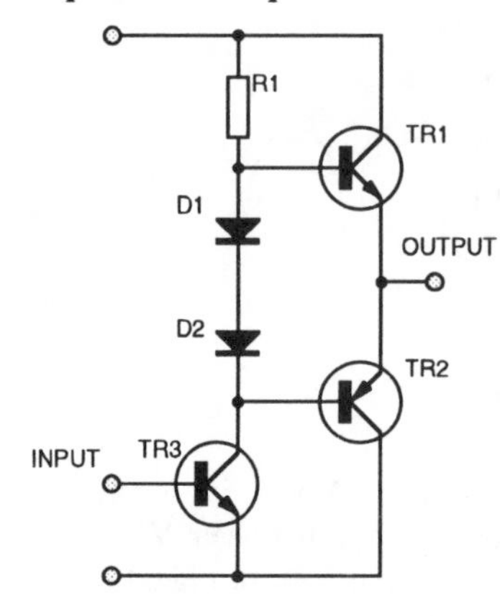

The circuit diagram shows a class A-B push-pull audio power amplifier. If D1 and D2 were not present, there would be crossover distortion in the output stage TR1 and TR2 due to the fact that the base-emitter voltages of TR1 and TR2 have to be overcome before they can be turned on to supply current to the output. D1 and D2 cause TR1 and TR2 to be biased on slightly so that there is a small current flow through both devices from the positive rail to 0V. When a signal is being amplified, D1 and D2 thus ensure there is a smooth transition of conduction between TR1 and TR2. When TR1 and TR2 heat up, however, their base-emitter voltage-drops reduce which would lead to an increase in the current through them and a risk of overheating. If D1 and D2 are in thermal contact with TR1 and TR2, however, the forward drops in D1 and D2 will also reduce and tend to compensate for TR1 and TR2.

Catching inductive current

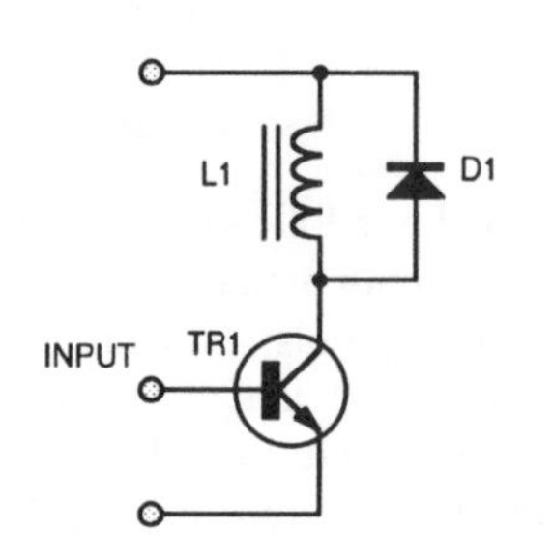

When a transistor is used to energize a relay or solenoid, energy is stored in the magnetic field of the solenoid due to the current flowing through it. When the transistor is turned off, the inductive current must continue to flow. If D1 was not present, then the voltage at the collector of TR1 would rise until the transistor was destroyed. The inductive current would then flow through the destroyed transistor until all the energy in the coil was dissipated. D1 provides a path for the inductive current which circulates through the diode which clamps the collector voltage of the transistor to approximately 0.6V above the positive supply rail.

Increasing the voltage withstand capability of another device

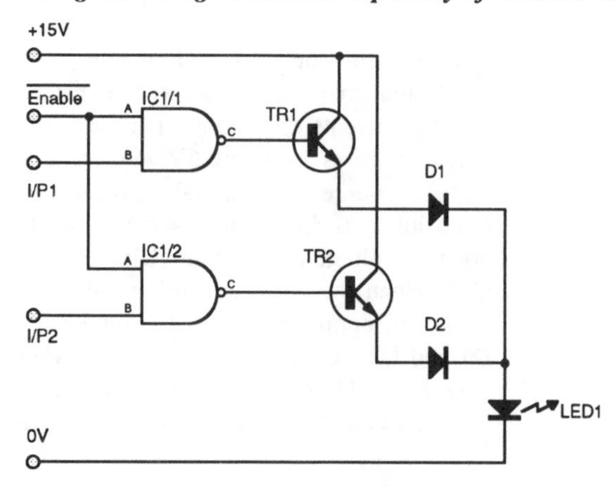

The circuit opposite uses TR1 and TR2 as emitter followers to increase the current drive of the NAND gates so as to drive LED1. The reverse breakdown voltage of the base-emitter junction of most transistors is only a few volts. The NAND gates are powered from 15V, consequently, if the output of IC1/1 was at a logic 0 and IC1/2 at a logic 1, the base-emitter junction of TR1 would break down if D1 was not present.

DC restoration

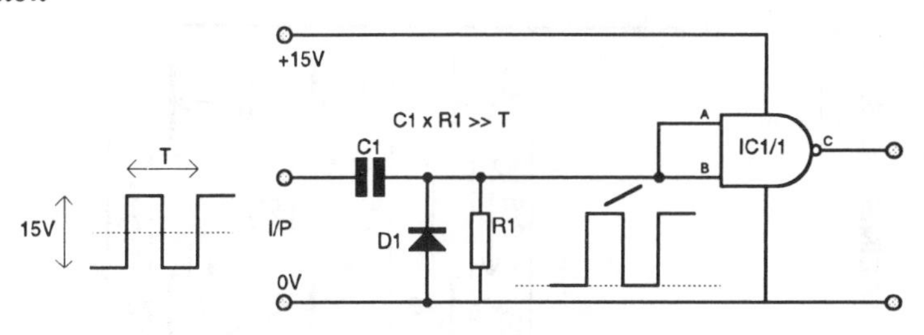

The square wave ac input to the above circuit has a peak to peak amplitude of 15V. It is required to drive a CMOS logic gate IC1/1 which is powered from a 15V rail. The input to the logic gate must ideally switch between 0V and +15V. If a diode is used to rectify the input waveform then the input to the logic gate would only switch from between approximately 0V and 7V. In the circuit above, C1 and D1 "pin" the input waveform's negative edge to the 0V rail so that almost the full 15V amplitude is now presented to the logic gate.

Testing a diode

A silicon diode is best tested with an analogue multimeter (such as an AVO). The meter has an internal battery which applies voltage to the diode to test whether it will conduct or block current. The positive of the battery is actually connected to the negative lead of the meter. This is because current normally flows into the meter when it is used on the volts and current ranges. When the meter is used on the ohms range, current flows from the meter to the component being tested; the meter needle would move in the wrong direction if the meter's internal battery was not connected with its positive terminal to the negative lead of the battery. (On digital multimeters, the positive lead of the internal battery is connected to the positive lead of the meter).

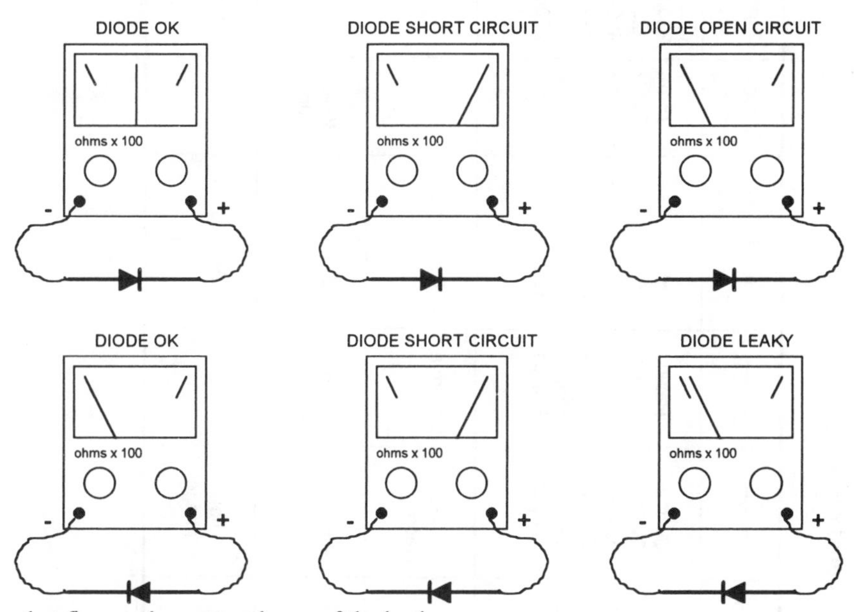

Ensure that fingers do not touch any of the leads.

Common signal diodes

TYPE No	TYPE	MANFR	V_R max	I_F (AV)	$I_{F\,(DC)}$ max	I_{FRM} max	I_{FSM} max	I_R max @ 25 °C	V_F max @ I_F	t_{rr} max	MAX DIMENSIONS
1N914	Switch	Philips	75V	-	75mA	225mA	0.5A, 1s	5uA	1V @ 10mA	4ns	
1N916	Switch	Philips	75V	-	75mA	225mA	0.5A, 1s	5uA	1V @ 10mA	4ns	
BAW62	Logic	Philips	75V	-	200mA	450mA	2A, 1us	5uA	1V@ 100mA	4ns	
1N4148	Switch	Philips	75V	-	200mA	450mA	2A, 1us	25nA*	1V @ 10mA	4ns	
1N4448	Switch	Philips	75V	-	200mA	450mA	2A, 1us	25nA*	1V @ 20mA	4ns	
OA202	Genpur	Philips	150V	-	160mA	250mA	---	100nA	1.15V@30mA	3.5us	
BY206	Fast Recovery	Motorola	350V	0.6A	-	-	25A	-	1.2V @ 2A	200ns	3.05, 6.6, 0.86
BA481	Schottky-barrier	Philips	4V	-	30mA	-	-	10uA	0.6 @ 10mA	-	1.7, 2.6, 0.55 DO34
BAT85	Schottky-barrier	Philips	30V		200mA	-	600mA	2uA	0.4V @ 10mA	5ns	
BYX10	High Voltage	G I	800V	0.3A	-	15A	-	1uA	1.6V @ 2A	2us typ	2.7, 5.2, 0.86 DO41

* @100 DEGREES C

Source: Motorola, Philips, GI

Fast-recovery diodes

In high frequency switching applications diodes are often used in situations where they are made to go from the forward conduction condition to the reverse blocking condition in a short period of time. If a diode is conducting, and then the voltage across it is quickly reversed, the diode will continue to conduct current (in the reverse direction) for a period of time until the diode blocks the flow of current. This period is called the reverse recovery time. For ordinary diodes, used for rectification at mains frequencies, the reverse recovery time may be several microseconds. The reverse recovery time for a fast diode is in the order of 50 to 150ns. If a normal slow diode is used in an application which requires a fast-recovery diode, then the diode and associated components may over-heat or be destroyed.

Uses of a fast-recovery diode

Dc to dc converter

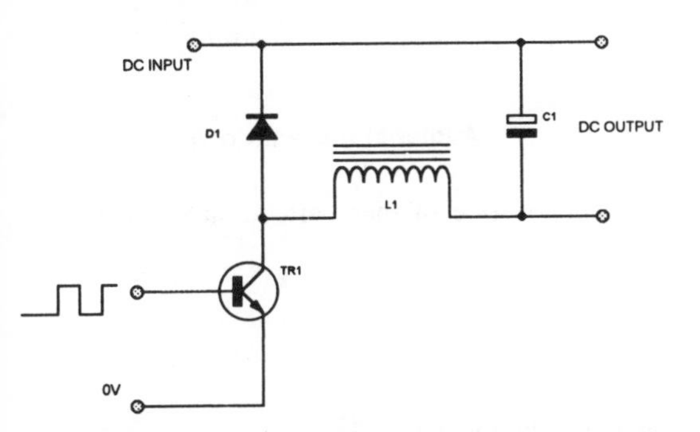

The circuit opposite is a standard dc to dc switching converter. The NPN transistor is turned on and off at high frequency and the ratio of the on to off times dictates the output dc voltage. When the transistor is on, current builds up in the inductor, L1. When the transistor turns off, the stored energy in the inductor causes the current through it to continue flowing through the diode D1 and the output capacitor C1 is charged up. When the transistor is turned on again, the voltage across D1 is suddenly reversed. Because of the reverse recovery phenomenon, the diode continues to conduct current but this time in the reverse direction until it recovers. Consequently there is a direct path for current to flow from the dc input supply, through the diode, and through the transistor to 0V. A high pulse of current flows until the diode reverts to the blocking condition. If a normal rectifier diode is used for D1, then both the transistor and the diode may over-heat and destroy themselves. A fast-recovery diode must be used for D1 so as to reduce the time period that the high current pulse flows and consequently reduce the dissipation in both devices.

Zener diodes

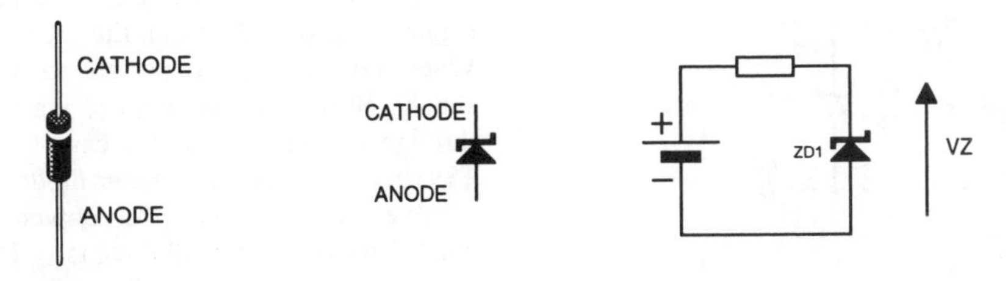

If a diode manufactured from heavily doped silicon is reverse biased, reverse conduction gradually starts to occur at around 6V. This is termed zener breakdown and is also seen on reverse biased base-emitter junctions of transistors. If the silicon is lightly doped however, reverse conduction occurs at a much higher reverse voltage and the effect happens more suddenly. This is termed avalanche breakdown.

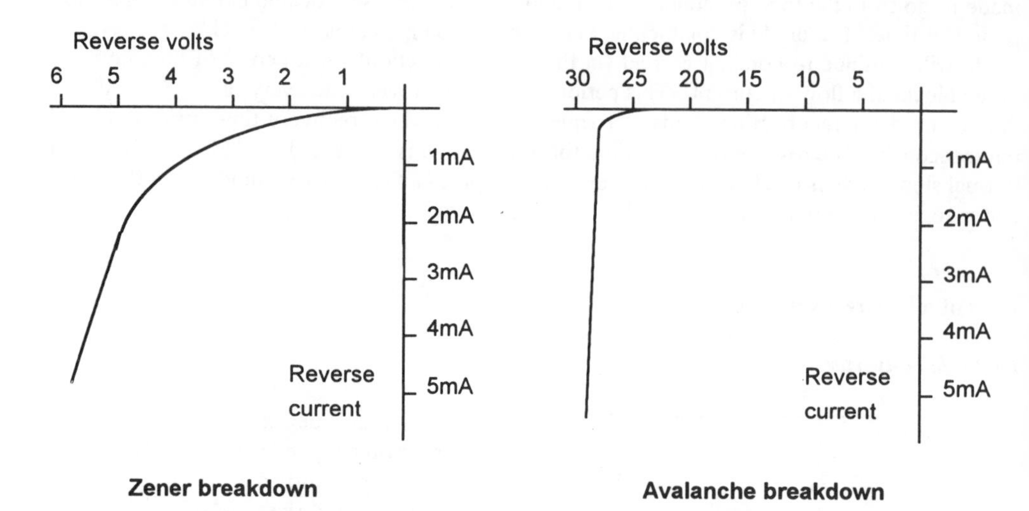

Zener breakdown **Avalanche breakdown**

Diodes which are specifically manufactured to employ one of these effects are called zener diodes.

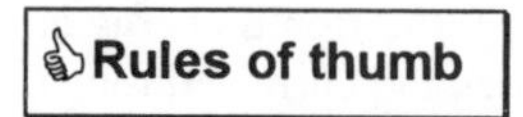

- The current through a zener diode must be limited by the external circuit.
- Zeners below about 5.1V have a "soft" characteristic - zener breakdown.
- Zeners below about 5.1V have a negative temperature coefficient.
- Higher voltage zeners diodes have a sharper characteristic - avalanche breakdown.
- Zeners above about 5.1V have a positive temperature coefficient.
- Zeners around about 5.1V have almost zero temperature coefficient.
- Zeners may also be used in the normal forward diode mode.

Uses of a zener diode

Voltage regulation

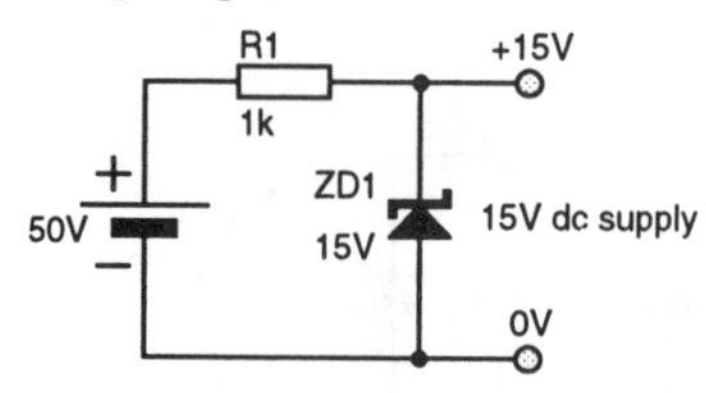

In the example opposite, a 15V dc regulated supply is produced from the 50V battery. When no load is connected to the 15V supply, all the current supplied via R1 flows through the zener. In this case it is (50 - 15)/1000 = 35mA. The zener diode must be capable of dissipating the power in this condition which is 0.525 watts. The 15V supply is only capable of supplying about 30mA after which the output voltage starts to fall.

Threshold device

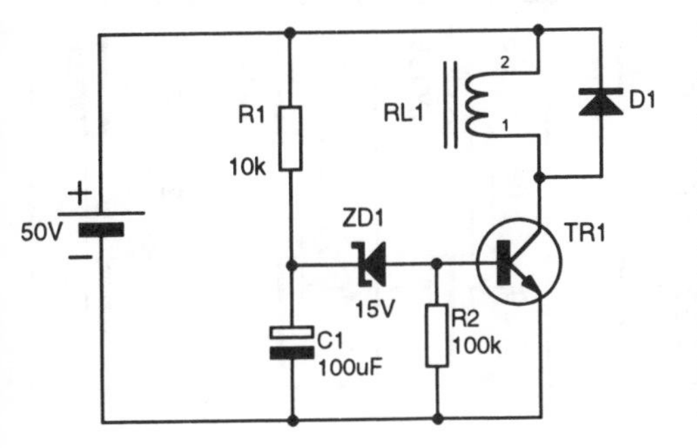

When power is first applied, the voltage across C1 (and hence across ZD1) is zero and the zener diode cannot conduct. C1 charges up via R1 and when the voltage exceeds 15V, ZD1 starts to conduct, when there is sufficient current flowing through ZD1 and into the base of TR1, the transistor turns on and energizes the relay. The circuit thus provides a time delay for the turn on of a relay when power is first applied.

Voltage clamp

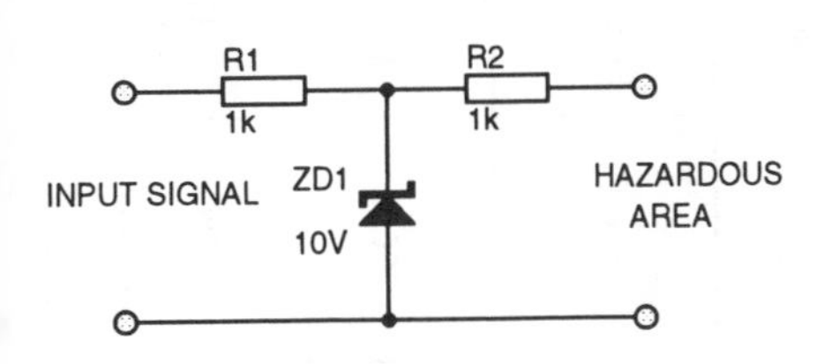

Electronic circuits are sometimes required to operate in hazardous environments where explosive gases may be present. It is essential to limit the electrical power reaching those circuits to prevent ignition of the gas. The circuit opposite is referred to as a zener barrier; the zener diode ZD1 limits the maximum voltage to 10V. The current through the zener is limited by R1 and the current to the hazardous area is limited by R1 and R2.

Testing a zener diode

Is the same procedure as testing a diode. See p. 65.

Zener diodes 500mW

WATTAGE	TYPE No	RANGE	TOL AVAIL	Tj °C MAX	MANFR		MAX DIMENSIONS
500mW	BZV55	2V7 to 20V	±2% to ± 20%	175°C	Thomson		1.7, 3.04, 0.55 DO34
500mW	BZV60	2V4 to 75V E24	±2%, ±5%	-	Philips		1.7, 3.04, 0.55 DO34
500mW	BZX55	0V8 to 62V 2V4 to 91V	±2%, ±5%	175°C 200°C	ITT Thomson Motorola		2.29, 5.08, 0.56 DO35
500mW	BZX79	2V4 to 75V 2V4 to 200V	±2%,±5%	155°C	Philips Motorola		2.29, 5.08, 0.56 DO35
500mW	BZX83	2V7 to 33V	±5%	175°C	Motorola		2.29, 5.08, 0.56 DO35
500mW	BZY88	2V7 to 33V	±2%, ±5%	200°C	Motorola		2.71, 5.2, 0.86 DO41
500mW	ZPD M-ZPD	1V to 51V 3V9 to 100V	±2%, ±5%	175°C 200°C	ITT Motorola	ITT is DO35 outline	2.71, 5.2, 0.86 DO41

Zener diodes 1.3W

WATTAGE	TYPE No	RANGE	TOL AVAIL	Tj °C MAX	MANFR	$\theta_{junc\text{-}case}$	MAX DIMENSIONS
1.3W	BZD23	7V5 to 270	±5%	200°C	Philips	Not quoted	2.15, 4.3, 0.81 SOD81
1.3W	BZV85	3V6 to 75V	±5%	200°C	Philips	110°C / W	2.6, 4.8, 0.81 DO41
1.3W	BZX85	2V7 to 150 3V3 to 100	±2%,±5%	175°C	ITT, Thomson Motorola	110°C / W	2.71, 5.2, 0.86 DO41
1.3W	MZD	3V9 to200V	±1,2,5%	200°C	Motorola	Not quoted	2.71, 5.2, 0.86 DO41
1.3W	ZPY M-ZPY	3V9 to 100	±2%, ±5%	200°C	Thomson Motorola	110°C / W	2.71, 5.2, 0.86 DO41

Zener diodes 1.5 to 2W

WATTAGE	TYPE No	RANGE	TOL AVAIL	Tj °C MAX	MANFR	$\theta_{junc\text{-}case}$	MAX DIMENSIONS
1.5W	BZY97	3V3 to 200 7V5 to 200	±5%	150°C	Thomson Fagor	60°C / W	3.05 6.35 0.86 F126
1.5W	ZGP10	100 to 200V	±5,±10,± 20%	175°C	G I	Not quoted	2.71 5.2 0.86 DO41

Light-emitting diodes

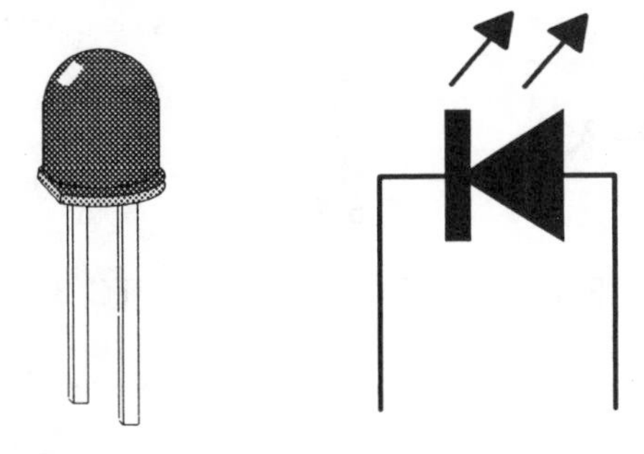

When certain p-n semiconductor junctions are forward biased, some charge carriers recombine and this is accompanied by the emission of visible light. The amount of light tends to be proportional to the forward current flowing through the LED. The colours commonly available are red, orange, yellow and green although red is the most efficient.

Rules of thumb

- An LED is a current-operated device which is suited to low supply voltage systems.
- An LED has a forward voltage drop = 1.5 to 2V.
- The reverse breakdown voltage of an LED is very low at only a few volts.
- The body of an LED has a flat section adjacent to one of the leads. In order to remember which way to connect an LED: current **F**lows to the **F**lat.

Uses of an LED

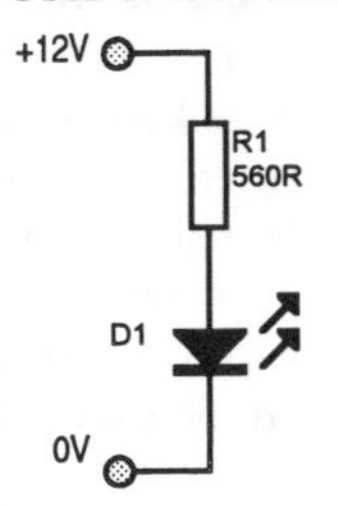

The LED is mainly used as an indicating device. Unlike a filament lamp however, the LED is not able to control its own current when connected to a voltage supply. The current through the LED must be set by external means and this is usually achieved by a series resistor. Operating currents of LEDs tend to be in the region 5 to 30mA. The series resistor may be calculated as follows:

$$R1 = (V_S - V_{LED}) / I_{LED}$$

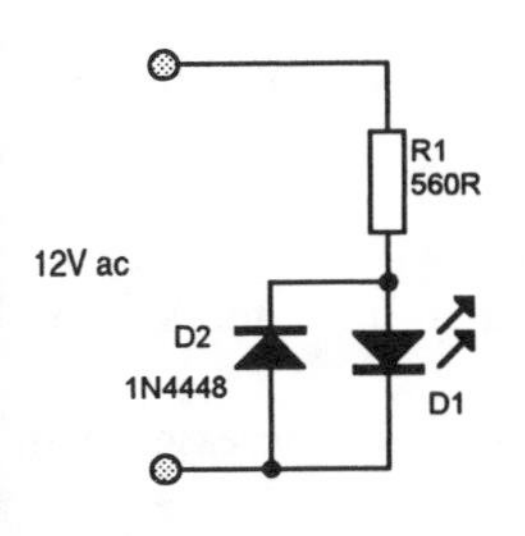

If an LED is to be used on an alternating supply then it must be protected from reverse voltages as typically an LED can only withstand a few volts. This may be achieved by placing a diode in reverse parallel with the LED. The series resistor may have to be reduced in value as the LED will only be conducting on positive half cycles of the supply and consequently will not appear as bright as in the 12V dc circuit. The maximum peak current through the LED must not exceed the LED's data sheet maximum value, however.

Bipolar transistors

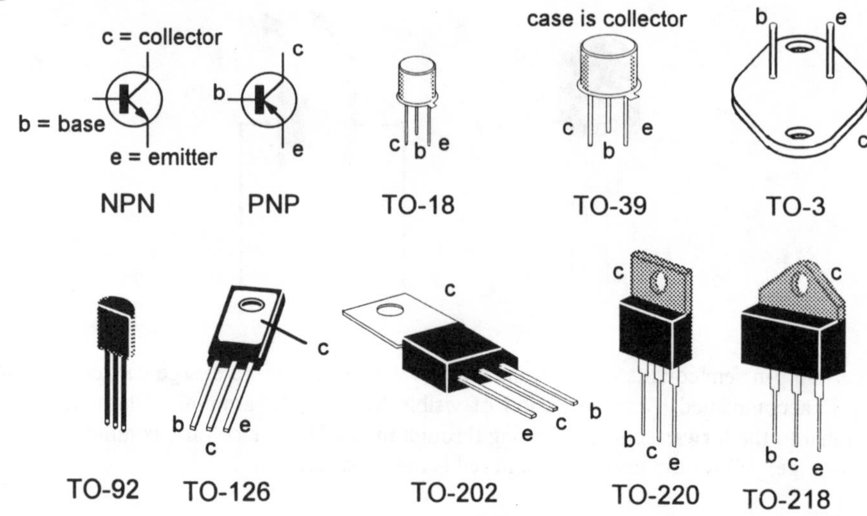

The bipolar transistor is a current-operated amplifying device. When a current is injected into the base with respect to the emitter, a much larger current is caused to flow between the collector and the emitter and may be in the order of 100 times the magnitude of the base current. Although the collector to emitter current is slightly affected by the voltage applied between the collector and the emitter, the current may be considered a source which is controlled by the base current.

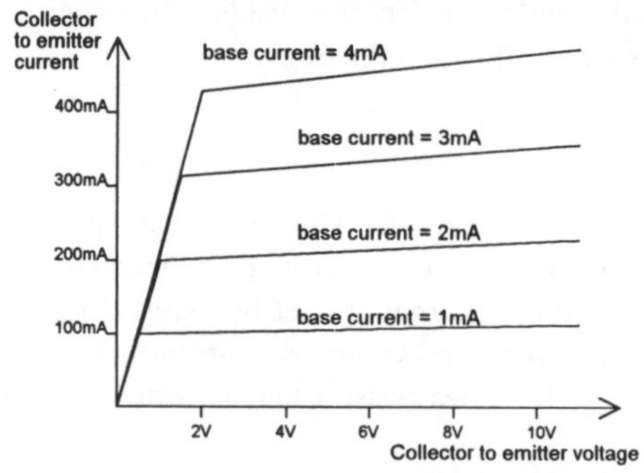

The curves opposite show the collect to emitter current for a typical NPN transistor for four different values of injected base current. When 1 mA is injected into the base with respect to the emitter, then provided the voltage between the collector and the emitter is greater than approximately 0.5V, 100mA is caused to flow between the collector and the emitter. As the collector to emitter voltage is increased, the collector to emitter current is also increased slightly and this is shown by the curve of "base current = 1mA" not being parallel to the x-axis.

Rules of thumb

- The base current flows i**N** on an **NPN** (out on a PNP).
- When current is injected into the base, a voltage drop of approx 0.6V is created between the base and emitter.
- This voltage drop reduces by 2.5mV for every ^{o}C rise in temperature.
- A reverse biased base-emitter junction zeners at about 5V.
- The collector to emitter voltage saturation voltage is approximately 0.3V for a signal transistor. It may be a volt or more for a power transistor.
- The current gain between the base -emitter and the collector -emitter increases with increase in temperature.
- The collector -emitter leakage current doubles for every 10 ^{o}C rise in temperature.

Uses of a transistor

Common- emitter voltage amplifier

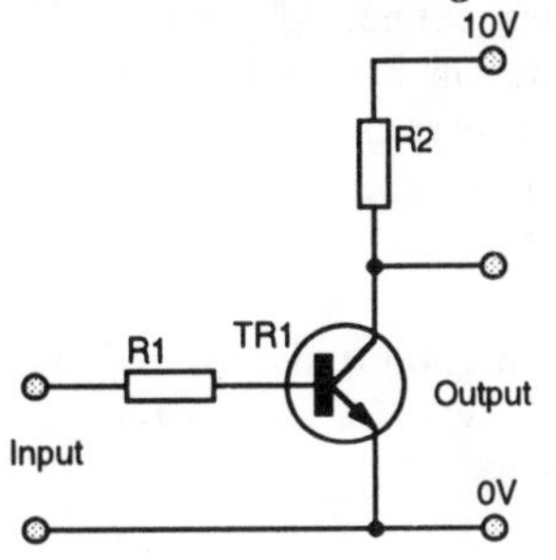

Although the bipolar transistor is a current amplifying device, it may be configured as a voltage amplifier by using a resistor (R1) to convert the input voltage signal to a current suitable for driving the base and using a resistor (R2) to convert the collector current of the transistor to an output voltage. The input signal voltage must be greater than the base- emitter voltage of 0.6V, however. There is an inversion from input to output in that the output voltage measured with respect to 0V falls as the input voltage rises.

Current source

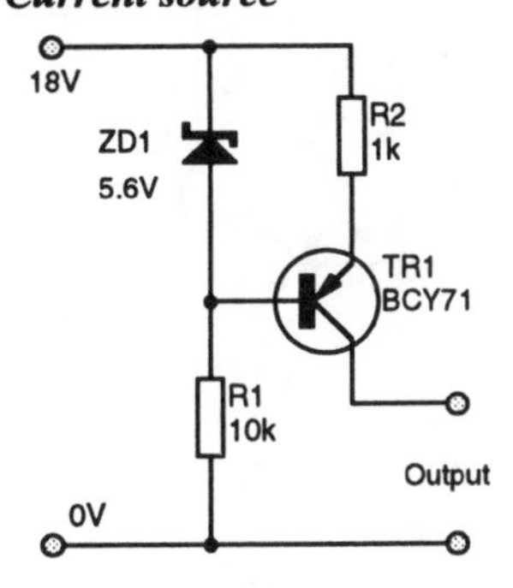

The base of TR1 is set at 5.6V below the 18V rail by ZD1. Consequently the emitter is set at approximately 5V below the 18V rail. A current of 5mA therefore flows through R2. Most of this current continues to flow out of the collector of TR1 and the collector constitutes a constant current source which may drive a load of between a short circuit and 10k without the current being significantly affected.

Transistor switch

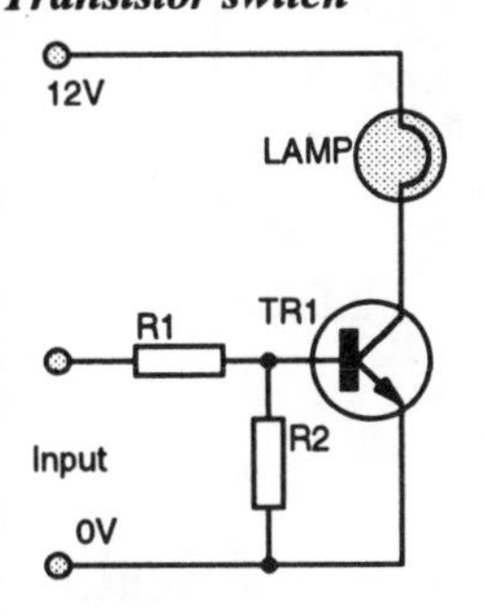

A transistor may be used as a switch having only two possible states; on or off. It may be used to turn on a lamp or other device in response to a signal applied to the base of the transistor. In the circuit opposite, a 12V lamp is being controlled by TR1. When a dc signal is applied to the input, TR1 is turned on by the base current derived by R1. The value of R1 must be such that the base current times the current gain of TR1 exceeds the current rating of the lamp. This ensures that TR1 saturates i.e. the collector -emitter voltage is almost zero. The lamp then receives its rated voltage. When the input signal is removed, the transistor turns off and the lamp is extinguished. R2 ensures the transistor cannot be turned partially on by providing an alternative path for the collector-base leakage current.

Testing a transistor

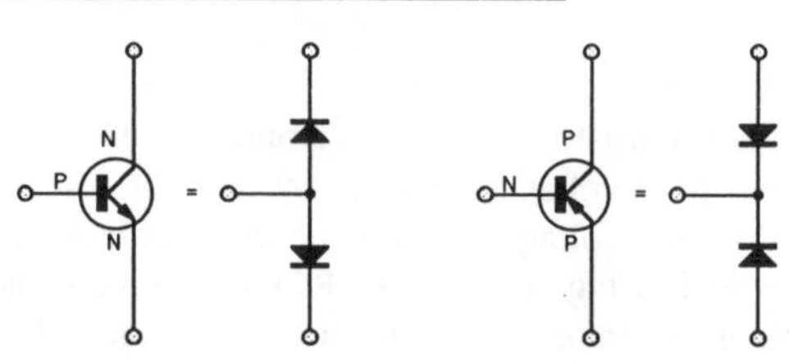

This is a similar process to testing a diode. See p. 65. The transistor may be considered to be two series connected diodes which have their anodes in common at the base terminal for an NPN transistor (the cathodes are in common at the base for a PNP transistor).

It is a good idea to familiarize yourself with the readings of a known good transistor and make a mark on the meter's scale with a felt-tip pen as a reference. The forward biased junctions of a power transistor give a greater deflection than a signal transistor.

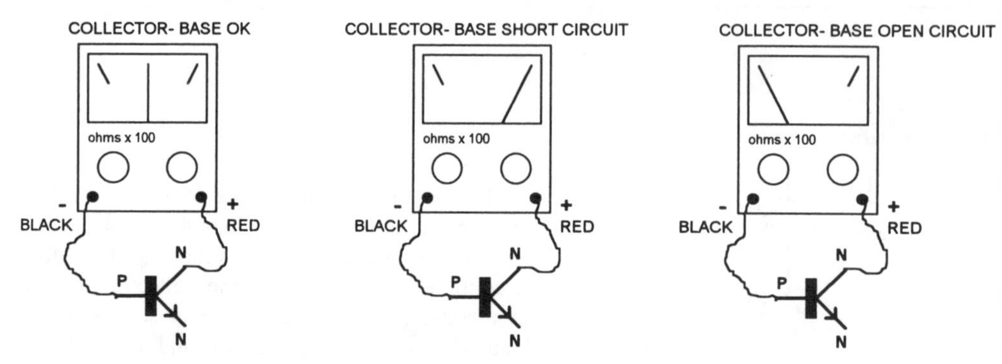

TESTING THE COLLECTOR- BASE JUNCTION WITH FORWARD BIAS

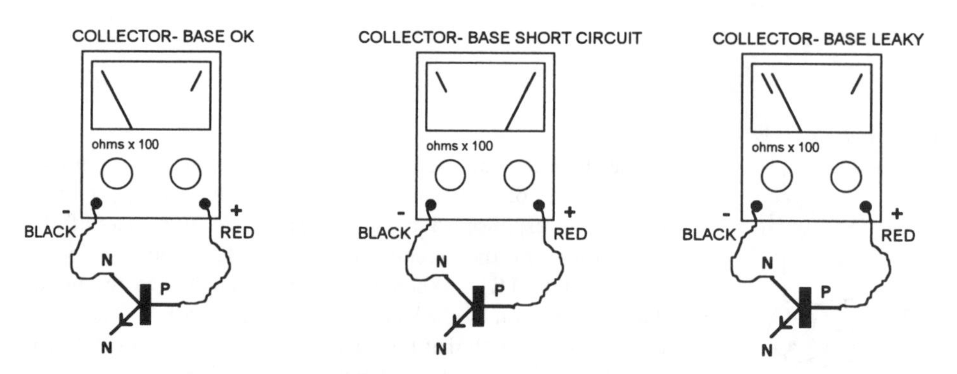

TESTING THE COLLECTOR- BASE JUNCTION WITH REVERSE BIAS

All three combinations of leads must be tested, i.e. collector- base, emitter- base, and collector-emitter. The above procedure is repeated for the emitter- base junction. The collector- emitter test should give no deflection in either direction (even with the meter set to ohms x 10k) unless the transistor is a special type with an integral collector to emitter diode e.g. a television line-output transistor. Care must be taken not to touch the test leads as leakage through the body can be mistaken for a leaky junction. Testing of a PNP transistor is simply the inverse of the procedure detailed above (swap the meter leads).

Bipolar transistor parameter definitions

Ic cont (maximum)

The maximum continuous dc collector current which may be drawn, provided:

1. due regard is given to the safe operating area curve published by the transistor manufacturer,
2. the transistor is not made to dissipate power which would cause the junction temperature to exceed Tj.

V_{CEO}

The maximum allowable collector to emitter voltage with the base terminal left open circuit. This is generally the worst case condition such that all other operating modes will be safe if the designer does not allow the collector to emitter voltage to exceed this value.

P_{tot} (max)

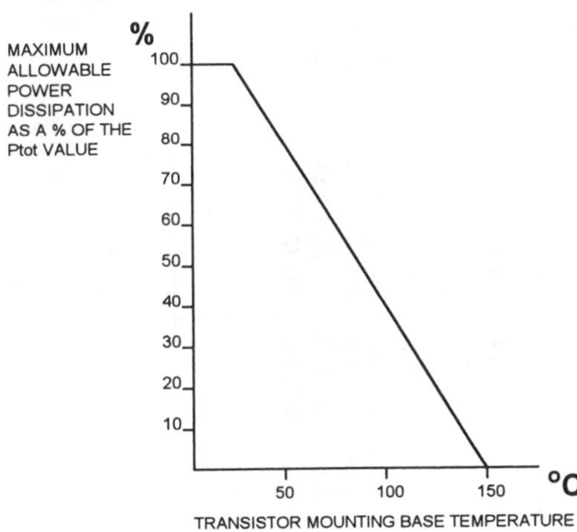

Ptot is the maximum allowable power dissipation (in watts) when the mounting base of the transistor is at 25°C . This is very rarely achievable, and the maximum allowable power of the transistor must be derated for mounting base temperatures of greater than 25°C according to a de-rating chart. A typical chart is shown opposite.

H_{FE} (min) @ I_C

This is the ratio of dc collector current to dc base current at the stated collector current. The figure given is the guaranteed minimum value at 25 °C. H_{FE} increases in value as temperature increases, and reduces in value as collector current increases.

Tj °C

This is the maximum allowable junction temperature in degrees C (see Ptot above).

ICEO max

This is the collector cut-off current with the base open circuit. It is a measure of the maximum leakage current which would flow when the collector-base junction is reverse biased (which is the normal condition). It is given at 25 °C. It tends to double in value for every 10 °C increase in temperature.

f_T (min)

This is called the transition frequency. It is the frequency at which the small signal gain of the transistor tends to extrapolate to unity. It is a measure of the transistor's high frequency performance.

°C/W junc - case

This is the thermal impedance of the transistor and represents the rise in temperature (for each watt of power dissipated) of the transistor's chip above the temperature of the transistor's metal mounting surface. It is used to calculate the chip temperature when the transistor is mounted on a heat sink.

Common TO92 transistors

OUTLINE	NPN	PNP	MANFR	APPL	V_{CEO}	I_C max	P tot @25°C	h_{FE} min @ I_C	Tj °C	I_{CBO} max @ 25°C	f_T (typ)	°C /W Junc-amb
TO92 (c b e)	BC237		Motorola	Amplifier	45V	0.1A	0.35W	120 @ 2mA	150°C	15nA	100MHz	357
		BC307			-45V	-0.1A	0.35W	120 @ 2mA	150°C	15nA	100MHz	357
	BC546		Philips		65V	0.1A	0.5W	110 @ 2mA	150°C	15nA	300MHz	250
		BC556			-65V	-0.1A	0.5W	75 @ 2mA	150°C	15nA	300MHz	250
	BC547				45V	0.1A	0.5W	110 @ 2mA	150°C	15nA	300MHz	250
		BC557			-45V	-0.1A	0.5W	75 @ 2mA	150°C	15nA	300MHz	250
	BC548				30V	0.1A	0.5W	110 @ 2mA	150°C	15nA	300MHz	250
		BC558			-30V	-0.1A	0.5W	75 @ 2mA	150°C	15nA	300MHz	250
	BC549			Low-	30V	0.1A	0.5W	125 @2mA	150°C	15nA	300MHz	250
		BC559		noise	-30V	-0.1A	0.5W	125 @2mA	150°C	15nA	300MHz	250
	BC337			Audio	45V	0.5A	0.8W	100 @ 0.1A	150°C	15nA	200MHz	200
		BC327		driver	-45V	-0.5A	0.8W	100 @ 0.1A	150°C	100nA	200MHz	200
TO92 (e b c)	2N3903		Philips	Switching	40V	0.2A	0.625W	50 @ 10mA	150°C		250MHz	357
		2N3905			-40V	-0.2A	0.625W	50 @ 10mA	150°C		200MHz	357
	2N3904				65V	0.1A	0.625W	100 @ 10mA	150°C		300MHz	357
		2N3906			-65V	-0.1A	0.625W	100 @ 10mA	150°C		250MHz	357
	MPSA43			High	200V	0.5A	0.625W	40 @ 10mA	150°C		50MHz	200
		MPSA93		voltage	-200V	-0.5A	0.625W	40 @ 10mA	150°C		50MHz	200
	MPSA42				300V	0.5A	0.625W	40 @ 10mA	150°C		50MHz	200
		MPSA92			-300V	-0.5A	0.625W	40 @ 10mA	150°C		50MHz	200

Common TO92 transistors - continued

OUTLINE	NPN	PNP	MANFR	APPL	V_{CEO}	I_C max	P tot @25°C	h_{FE} min @ I_C	Tj °C	I_{CBO} max @ 25 °C	f_T (typ)	°C /W Junc-amb
TO92	BC182L		Motorola	General purpose	50V	0.1A	0.35W	100 @ 2mA	150°C	15nA	150MHz	357
e c b		BC212L	Philips		-50V	-0.1A	0.35W	60 @ 2mA	150°C	15nA	150MHz	357
	BC183L				30V	0.1A	0.35W	100 @ 2mA	150°C	15nA	150MHz	357
		BC213L			-30V	-0.1A	0.35W	80 @ 2mA	150°C	15nA	150MHz	357
	BC184L			Low noise	30V	0.1A	0.35W	250 @ 2mA	150°C	15nA	150MHz	357
		BC214L			-30V	-0.1A	0.35W	140 @ 2mA	150°C	15nA	150MHz	357
	2N3704			General purpose	30V	0.8A	0.35W	100 @50mA	150°C	15nA	150MHz	357
		2N3702			-25V	-0.2A	0.35W	100 @50mA	150°C	15nA	150MHz	357
	2N3705				30V	0.8A	0.62W	250 @2mA	150°C	15nA	150MHz	357
		2N3703			-30V	-0.1A	0.62W	140 @2mA	150°C	15nA	150MHz	357
	TIPP31			Power switching	30V	2A	0.8W	20 @ 1A	150°C	15nA	150MHz	357
		TIPP32			-30V	-2A	1.25W	25 @ 1A	150°C	15nA	150MHz	357
	TIPP31A				60V	2A	0.8W	20 @ 1A	150°C	15nA	150MHz	357
		TIPP32A			-60V	-2A	1.25W	25 @ 1A	150°C	15nA	150MHz	357
	TIPP31B				80V	2A	0.8W	20 @ 1A	150°C	15nA	150MHz	357
		TIPP32B			-80V	-2A	1.25W	25 @ 1A	150°C	15nA	150MHz	357
	TIPP31C				100V	2A	0.8W	20 @ 1A	150°C	15nA	150MHz	357
		TIPP32C			-100V	-2A	1.25W	25 @ 1A	150°C	15nA	150MHz	357
	BC639			Audio drivers	80V	1A	1W	40 @ 0.15A	150°C	100nA	130MHz	156
		BC640			-80V	-1A	1W	40 @ 0.15A	150°C	100nA	50MHz	156

Source: Motorola, Philips

Common TO39 transistors

OUTLINE	NPN	PNP	MANFR	APPL	V_{CEO}	I_C max	P tot @25°C	h_{FE} min @ I_C	Tj °C	I_{CBO} max @ 25 °C	f_T (typ)	°C /W Junc-case
TO39 c b e	BFY51		Philips	Gen pur	30V	1A	0.8W	40 @0.15A	200°C	500nA	50MHz	35
	BFY50		Philips	Gen pur	35V	1A	0.8W	30 @0.15A	200°C	500nA	60MHz	35
	2N2219		Philips	Switch	30V	0.8A	0.8W	75 @10mA	200°C	10nA	250MHz	50
		2N2905	Philips	Switch	-40V	-0.6A	0.6W	100@0.15A	200°C	20nA	200MHz	58
	2N3053		Philips	Amp	40V	0.7A	5W	50 @0.15A	200°C	250nA	100MHz	35
		2N4037	Harris	Gen pur	-40V	-1A	7W	50 @0.15A	200°C	250nA	60MHz	25
	2N1711		Philips	Amp	50V	1A	0.8W	100 @0.15A	200°C	10nA	70MHz	58.3
		2N4036	Philips	Amp	-65V	-1A	7W	20 @0.15A	200°C	20nA	?	25
	BC301		Thomson	?	60V	1A	0.85W	40 @0.15A	175°C	?	60MHz	?
	BFX85		Philips	Gen pur	60V	1A	0.8W	70 @0.15A	175°C	500nA	50MHz	35
	2N3019		Philips	Amp	80V	1A	0.8W	100 @0.15A	200°C	10nA	100MHz	35
	2N3440		Philips	Switch	250V	1A	1W	40 @20mA	200°C	100nA	70MHz	58.3
		2N5415	Philips	Switch	-200V	-1A	1W	30 @50mA	200°C	50uA	15MHz	17.5
	2N3439		Philips	Switch	350V	1A	1W	30 @2mA	200°C	100nA	70MHz	58.3
		2N5416	Philips	Switch	-300V	-1A	1W	30 @50mA	200°C	50uA	15MHz	17.5

Source:- Philips,Harris, Thomson

1A - TO220 power transistors

OUTLINE	NPN	PNP	MANFR	APPL	I_C cont	V_{CEO}	Ptot @25°C	h_{FE} min @ I_C	Tj °C	I_{CEO} max @ 25°C	f_T (min)	°C /W Junc-case
TO220	TIP29A		Motorola	Gen pur	1A	60V	30W	15 @1A	150°C	0.3mA	3MHz	4.167
		TIP30A		Gen pur	1A	-60V	30W	15 @1A	150°C	0.3mA	3MHz	4.167
	TIP29B			Gen pur	1A	80V	30W	15 @1A	150°C	0.3mA	3MHz	4.167
		TIP30B		Gen pur	1A	-80V	30W	15 @1A	150°C	0.3mA	3MHz	4.167
	TIP29C			Gen pur	1A	100V	30W	15 @1A	150°C	0.3mA	3MHz	4.167
		TIP30C		Gen pur	1A	-100V	30W	15 @1A	150°C	0.3mA	3MHz	4.167
	TIP47			Audio & switchg	1A	250V	40W	10 @1A	150°C	1mA	10MHz	3.125
	TIP48				1A	300V	40W	10 @1A	150°C	1mA	10MHz	3.125
		MJE5730			1A	-300V	40W	10 @1A	150°C	1mA	10MHz	3.125
	TIP49				1A	350V	40W	10 @1A	150°C	1mA	10MHz	3.125
		MJE5731			1A	-350V	40W	10 @1A	150°C	1mA	10MHz	3.125
	TIP50				1A	400V	40W	10 @1A	150°C	1mA	10MHz	3.125
		MJE5732			1A	-400V	40W	10 @1A	150°C	1mA	10MHz	3.125

Source:- Motorola

2 to 3A - TO220 power transistors

OUTLINE	NPN	PNP	MANFR	APPL	I_C cont	V_{CEO}	P tot @25°C	h_{FE} min @ I_C	Tj °C	I_{CEO} max @ 25 °C	f_T (min)	°C /W Junc-case
TO220 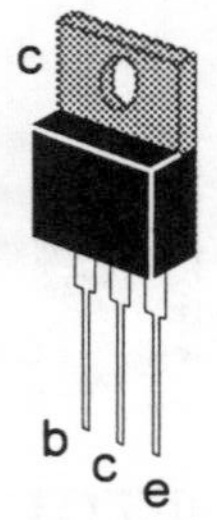	BD239A		Motorola	Gen pur	2A	60V	30W	15 @ 1A	150°C	0.3mA	3MHz	4.167
		BD240A			-2A	-60V	30W	15 @ 1A		0.3mA	3MHz	4.167
	BD239B				2A	80V	30W	15 @ 1A		0.3mA	3MHz	4.167
		BD240B			-2A	-80V	30W	15 @ 1A		0.3mA	3MHz	4.167
	BD239C				2A	100V	30W	15 @ 1A		0.3mA	3MHz	4.167
		BD240C			-2A	-100V	30W	15 @ 1A		0.3mA	3MHz	4.167
	BUX84			SMPS	2A	400V	50W	30 @ 0.1A		0.2mA Ices	4MHz	2.5
	BUX85			SMPS	2A	450V	50W	30 @ 0.1A		0.2mA Ices	4MHz	2.5
	BU505		Philips	Deflecn	2.5A	700V	75W	2.2 @ 2A		-	-	-
	TIP31A		Motorola	Gen pur	3A	60V	40W	10 @ 3A		0.3mA	3MHz	3.125
		TIP32A			-3A	-60V	40W	10 @ 3A		0.3mA	3MHz	3.125
	TIP31B				3A	80V	40W	10 @ 3A		0.3mA	3MHz	3.125
		TIP32B			-3A	-80V	40W	10 @ 3A		0.3mA	3MHz	3.125
	TIP31C				3A	100V	40W	10 @ 3A		0.3mA	3MHz	3.125
		TIP32C			-3A	-100V	40W	10 @ 3A		0.3mA	3MHz	3.125

Source:- Motorola & Philips

4 to 5A - TO220 power transistors

OUTLINE	NPN	PNP	MANFR	APPL	I_C cont	V_{CEO}	P tot @25°C	h_{FE} min @ I_C	Tj °C	I_{CEO} max @ 25 °C	f_T (min)	°C /W Junc-case
TO220	BD535		Motorola	Gen pur	4A	60V	50W	25 @ 2A	150	Ices 0.1mA	3MHz	2.5
		BD536			-4A	-60V	50W	25 @ 2A		0.1mA	3MHz	2.5
	D44C10			Switchg	4A	80V	30W	10 @ 2A		0.01mA	50 typ	4.2
		D45C10			-4A	-80V	30W	10 @ 2A		0.01mA	40 typ	4.2
	BUL52B		Semilab	SMPS	4A	400V						
	MJE13004		Motorola	SMPS	4A	300V	75W	8 @ 2A		Icev 1mA	4MHz	1.67
	MJE13005				4A	400V	75W	8 @ 2A		1mA	4MHz	1.67
	BUT11		Thomson	High voltage switch	5A	400V	83W	-		Ices 1mA	-	1.5
	BUV46		Thomson		5A	400V	70W	-		Icer 0.1mA	-	1.76

Source:- Motorola & Thomson

6 to 7A - TO220 power transistors

OUTLINE	NPN	PNP	MANFR	APPL	I_C cont	V_{CEO}	Ptot @25°C	h_{FE} min @ I_C	Tj °C	I_{CEO} max @ 25°C	f_T (min)	°C /W Junc-case
TO220	TIP41A		Motorola	Gen pur	6A	60V	65W	15 @ 3A	150	0.7mA	3MHz	1.92
		TIP42A			-6A	-60V	65W	15 @ 3A		0.7mA	3MHz	1.92
	TIP41B				6A	80V	65W	15 @ 3A		0.7mA	3MHz	1.92
		TIP42B			-6A	-80V	65W	15 @ 3A		0.7mA	3MHz	1.92
	TIP41C				6A	100V	65W	15 @ 3A		0.7mA	3MHz	1.92
		TIP42C			-6A	-100V	65W	15 @ 3A		0.7mA	3MHz	1.92
	BD243A				6A	60V	65W	15 @ 3A		0.7mA	3MHz	1.92
		BD244A			-6A	-60V	65W	15 @ 3A		0.7mA	3MHz	1.92
	BD243B				6A	80V	65W	15 @ 3A		0.7mA	3MHz	1.92
		BD244B			-6A	-80V	65W	15 @ 3A		0.7mA	3MHz	1.92
	BD243C				6A	100V	65W	15 @ 3A		0.7mA	3MHz	1.92
		BD244C			-6A	-100V	65W	15 @ 3A		0.7mA	3MHz	1.92
	2N6292				7A	70V	40W	30 @ 2A		1mA	4MHz	3.125
		2N6107			-7A	-70V	40W	30 @ 2A		1mA	10MHz	3.125
	BU407			Deflec	7A	150V	60W	30 @ 1.5A		Ices 5mA	10MHz	2.08
	BU406				7A	200V	60W	30 @ 1.5A		5mA	10MHz	2.08

Source:- Motorola

Common TO3 power transistors

OUTLINE	NPN	PNP	MANFR	APPL	V_{CEO}	I_C max	P tot @25°C	h_{FE} min @ I_C	Tj °C	I_{CEO} max @ 25 °C	f_T (min)	°C /W Junc-case
TO3	2N3771		Motorola	Linear	40V	30A	150W	15 @15A	200°C	10mA	0.2MHz	1.17
	2N3055			Gen pur	60V	15A	115W	20 @ 4A	200°C	0.7mA	2.5MHz	1.52
		MJ2955		Gen pur	-60V	-15A	115W	20 @ 4A	200°C	0.7mA	2.5MHz	1.52
	2N3772			Linear	60V	20A	150W	15 @10A	200°C	10mA	0.2MHz	1.17
b e	2N3716			Amp	80V	10A	150W	50 @ 1A	200°C	-	-	1.17
		2N3792		Amp	-80V	-10A	150W	50 @1A	200°C	-	-	1.17
	MJ802			Audio	90V	30A	200W	25 @ 7.5A	200°C	Icbo 1mA	2MHz	0.875
	2N3773			Audio	140V	16A	150W	15 @ 8A	200°C	10mA	-	1.17
		2N6609		Audio	-140V	-16A	150W	15 @ 8A	200°C	10mA	-	1.17
	MJ15003			Audio	140V	20A	250W	25 @ 5A	200°C	10mA	2MHz	0.7
c		MJ15004		Audio	140V	20A	250W	25 @ 5A	200°C	0.25mA	2MHz	0.7
	BUX48A			SMPS	450V	15A	175W	8 @ 8A	200°C	0.25mA	-	1
CASE IS	MJ16002A			SMPS	500V	5A	125W	5 @ 5A	200°C	-	-	1.4
COLLECTOR	MJ16006				450V	8A	150W	5 @ 8A	200°C	-	-	1.17
	MJ16010				450V	15A	175W	5 @ 15A	200°C	-	-	1
	BU208A			TV	700V	5A	12.5W	2.25 @ 4.5A	200°C	-	4MHz typ	1.6

Source:- Motorola

Field-effect transistors

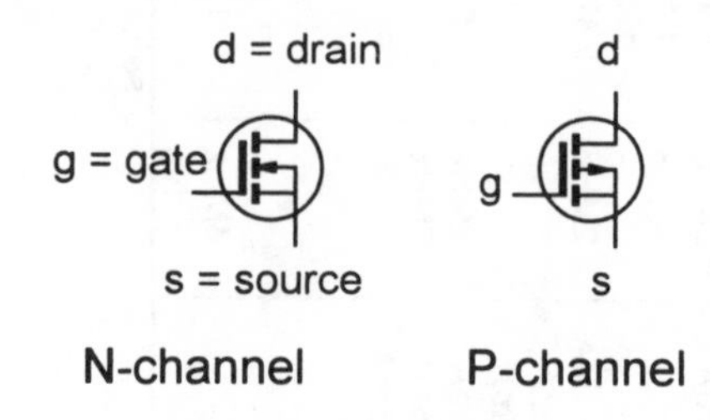

The fundamental difference between field-effect transistors and ordinary or bi-polar transistors is the method used to control the current through the transistor. In the bi-polar transistor, the current is controlled by varying the current into the base. In the field-effect transistor, the current is controlled by varying an electric field inside the transistor. This is achieved by varying the voltage applied to the control terminal (which is called the gate) with respect to the source terminal. The gate terminal has a very high impedance and draws very little current during normal operation. There are two main types of field-effect transistors (FETs), junction or JFETs and metal-oxide semiconductor or MOSFETs. Junction FETs are not used in industry to any significant degree as MOSFETs have much superior characteristics and are now available in many different ratings and for many different applications. This chapter will therefore concentrate on the characteristics and applications of MOSFETs.

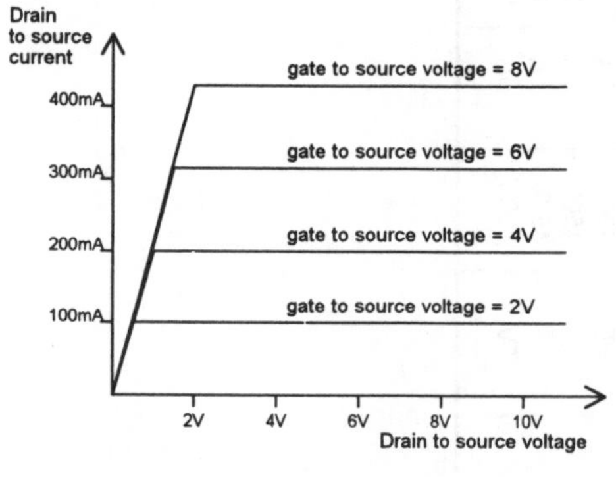

The curves opposite show the drain to source currents for four values of gate to source voltage for a typical N-channel MOSFET. When a voltage of 2V is applied between the gate and the source, then provided the voltage between the drain and the source is greater than about 0.5V, 100mA flows from the drain to the source. Notice that as the drain to source voltage is increased, the drain to source current remains constant. If the gate to source voltage is increased to 4V then the drain to source current now becomes 200mA. Again the current is not affected by the drain voltage. This means that a MOSFET is a true constant-current source device.

Rules of thumb

- A MOSFET is a voltage-operated device. The voltage on the gate (with respect to the source) controls the current from drain to source.
- There are two main types of MOSFET: N-channel and P-channel.
- On the drawing symbols, the arrow goes i**N** for **N**-channel.
- On an N-channel MOSFET the gate voltage is positive with respect to the source and the main current flows from the drain to the source. (**N**-channel is similar to an **N**PN bipolar transistor.) On a P-channel, everything is reversed.
- A MOSFET starts to turn on when the gate to source voltage is approximately 1 to 3V.
- The maximum gate to source voltage is usually about ±20V.

Uses of a MOSFET

Common- source voltage amplifier

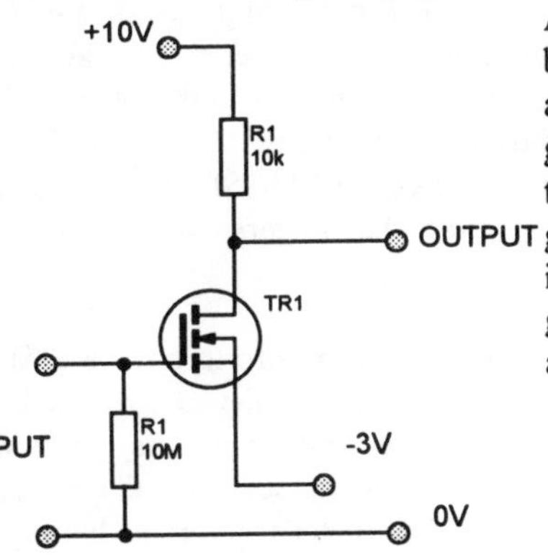

A MOSFET has a very high gate impedance and may be used to amplify signals from devices which are not able to supply much current. In the circuit shown, the gate to source is biased at 3V so that the MOSFET is turned on. A 10M resistor must be used to "tie" the gate to the 0V rail. This resistor actually sets the input impedance of the amplifier. It should be noted that the gain of a MOSFET amplifier of this type is not as high as the equivalent NPN bipolar transistor stage.

Current source

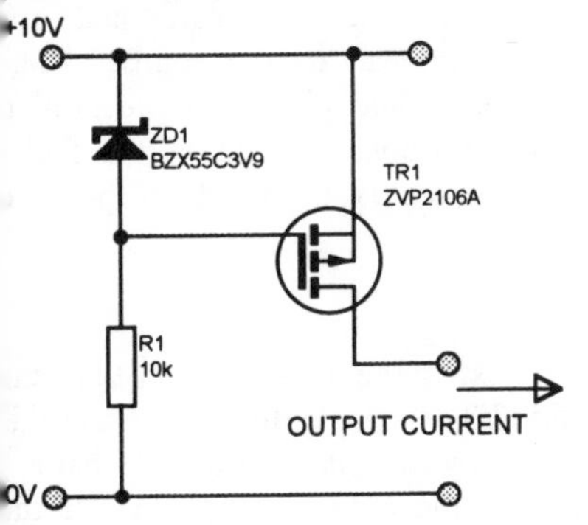

In the circuit opposite the P-channel MOSFET's gate is set at 3.9V negative with respect to its source. (The transistor is drawn upside-down.) A drain current of approximately 200mA flows out of the drain terminal.

Electronic switch

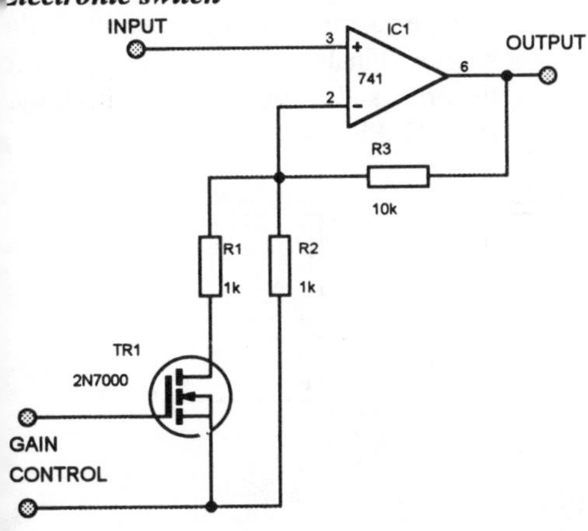

The circuit opposite is a dc amplifier whose gain may be changed by applying a dc voltage to the gate of the MOSFET. When 0V is applied to the gate, the MOSFET is off (open circuit) and the gain of the amplifier is:

$$1 + R3/R2 = 11$$

If approximately 10V is applied to the gate the MOSFET is turned hard on and acts like a switch connecting R1 to the 0V rail. The resistance between pin 2 of IC1 and 0V is now 500 ohms and the gain of the amplifier is increased to 21. It should be pointed out that the amplifier is only suitable for amplifying positive dc voltages.

Testing a MOSFET

A MOSFET may be tested using an analogue multimeter which has a high resistance range such as x10k. In order to measure high resistance values, an analogue meter is usually fitted with a 6V internal battery. The positive of the battery is actually connected to the negative lead of the meter (see "testing a diode"). This battery is used to turn the MOSFET on and off during testing. The test procedure described below is for testing an N-channel MOSFET. To test a P channel MOSFET, the meter leads must be reversed. Ensure that fingers do not touch the MOSFET leads during testing.

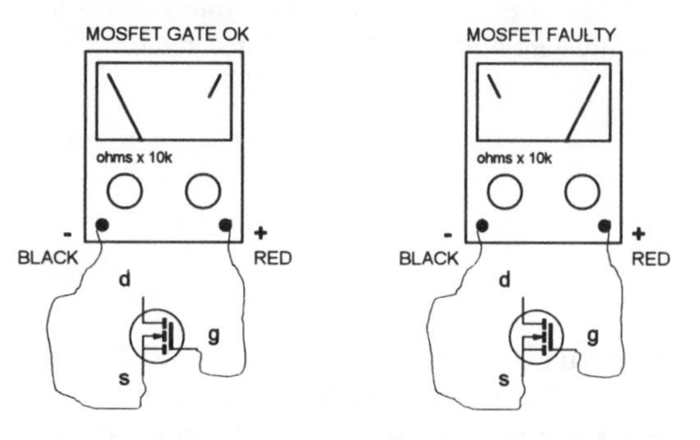

TURNING THE MOSFET OFF AND TESTING ITS GATE INSULATION

The first step is to turn the MOSFET off. This is achieved by applying a negative voltage to its gate with respect to its source. The negative of the meter's internal battery is connected to the positive lead, and so it is the positive lead which is connected to the gate and the negative lead to the source. The meter should give no deflection at all. If it does the MOSFET is faulty and there is no need to proceed further.

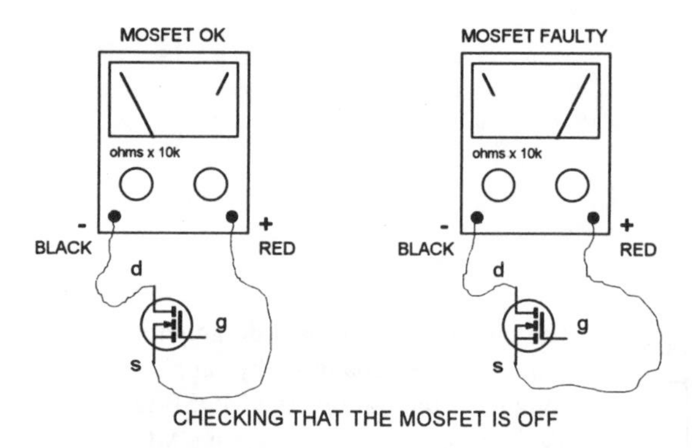

CHECKING THAT THE MOSFET IS OFF

The next stage is to check that the MOSFET is actually turned off and this is done by applying the meter's battery between the drain and source. If the meter shows any reading other than infinity then the MOSFET is faulty. The next stage is to turn the MOSFET on. This is done by applying a positive voltage to the gate with respect to the source. Finally the leads are applied to the drain and source again to check that the MOSFET is on.

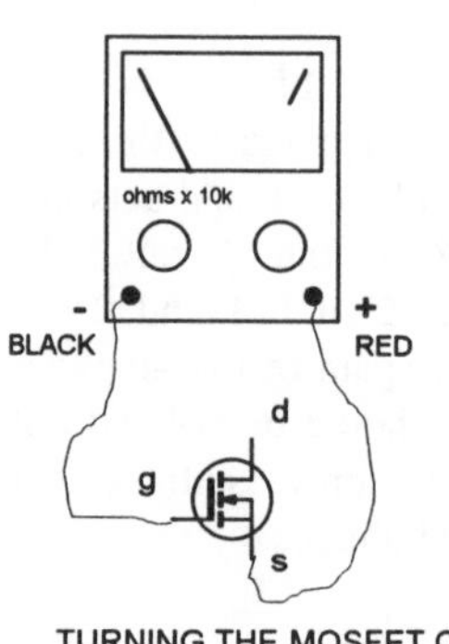

TURNING THE MOSFET ON

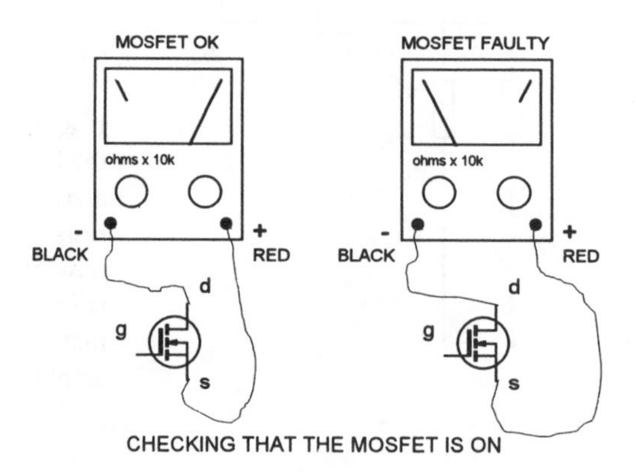

CHECKING THAT THE MOSFET IS ON

MOSFET parameter definitions

I_D max

The maximum continuous dc current which may be drawn from drain to source, provided:

1. due regard is given to the safe operating area curve published by the MOSFET manufacturer,
2. the MOSFET is not made to dissipate power which would cause the junction temperature to exceed Tj.

V_{DSS} max

This is the maximum drain to source voltage which may be applied (with the gate terminal short-circuited to the source).

V_{GS} max

This figure represents the maximum voltage which may be applied between the gate and the source. This is usually the same magnitude for either a positive or a negative voltage.

P_D max

P_D is the maximum allowable power dissipation (in watts) when the mounting base of the MOSFET is at 25°C . This is very rarely achievable, and the maximum allowable power of the MOSFET must be de-rated for mounting base temperatures of greater than 25°C according to a de-rating chart. A typical chart is shown in the section describing bipolar transistors.

V_{GS} (th)

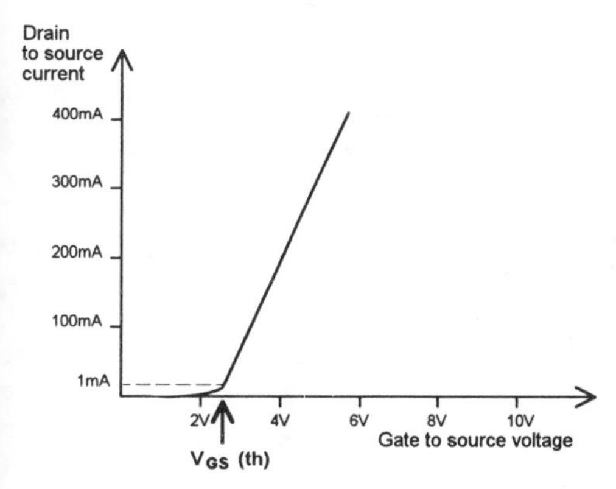

This is the gate to source threshold voltage at which the MOSFET is considered to be just turning on. Manufacturers usually quote a V_{GS}(th) for a drain current of 1mA. That is to say when a gate to source voltage of V_{GS}(th) is applied, the drain current is 1 mA. At gate to source voltages greater than V_{GS}(th), the drain current increases rapidly and then a linear region is reached where the drain current increases almost linearly with gate to source voltage as shown in the diagram.

g_{FS}

This is called the transconductance and is a measure of the gain of a MOSFET. It is the change in drain to source current for a change in gate to source voltage and is the slope of the linear region of the graph shown above. The transconductance is measured in siemens (S) which is amps per volt. This unit used to be referred to as the mho.

R_{DS} (on)

This is the drain to source resistance when the MOSFET is fully turned on. It is a measure of the device's power switching ability with regard to the watts dissipated by the MOSFET when it is in the on condition.

Low power MOSFETs

OUTLINE	N Channel	P Channel	MANFR	APPL	I_D cont	V_{DS}	P_D @25°C	r_{DS} (on) @ I_D	Tj °C	g_{fs} min @ I_D	$V_{GS(th)}$	°C /W Junc-amb
TO92 (s g d)	2N7000		Motorola	Gen pur	0.2A	60V	0.4W	5 ohm max @ 0.5A	150°C	100uS @ 0.2A	0.8V to 3V	312.5
	VN10KM		Siliconix	TO237	0.3A	60V	1W	5 ohm max @ 0.5A	150°C	100mS @ 0.5A	0.8 to 2.5V	125
		ZVP2106 A	Zetex	E line but same pin-out	-0.32A	-60V	0.7W	5 ohm max @ -0.5A			1.5V to 4.5V	
TO92 (d g s)	BSS98		Siemens		0.3A	50V	0.63W	3.5 ohm max @ 0.3A	150°C	120mS @ 0.3A	1.5 to 3.5V	150
	BST72A		Philips		0.3A	80V	0.83W	10 ohm max @ 0.15A	150°C	150mS @ 0.2A	0.5V to 1.5V	200
	BS107P		Ferranti		0.12A	200V	0.5W	30 ohm max @ 0.1A	150°C	360mS @ 0.4A		
		BSS110	Siemens	Logic	-0.17A	-50V	0.63W	10 ohm max @ -0.1A	150°C	50mS @ -0.1A	-0.8 to -1.8V	

TO220 power MOSFETs

OUTLINE	N Channel	P Channel	MANFR	I_D cont	V_{DS}	P_D @25°C	r_{DS} (on) @ I_D	Tj °C	g_{fs} min @ I_D	$V_{GS(th)}$ (typ)	°C /W Junc-case
TO220 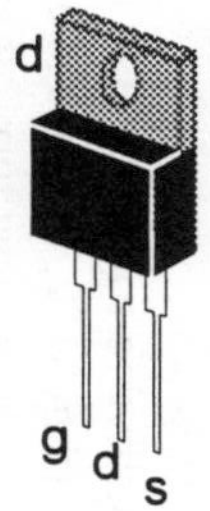	BUZ11		Siemens	30A	50V	75W	0.04R max @ 15A	150°C	4S @ 15A	2.1 to 4V	1.67 max
	IRF630		IR	9A	200V	74W	0.4R max @ 5.4A	150°C	3.8S @ 5.4A	2 to 4V	1.7 max
	IRF830		IR	4.5A	500V	75W	1.5R max @ 2.7A	150°C	2.5S @2.7A	2 to 4V	1.7 max
	IRF840		IR	8A	500V	125W	0.85R @ 8.5A	150°C	4.9S min @ 4.8A	2 to 4V	1 max
		IRF9520	IR	-6.8A	-100V	60W	0.6R max @ -4.1A	175°C	2S min @ -4.1A	-2 to -4V	2.5 max
		IRF9630	IR	-6.5A	-200V	74W	0.8R max @ -3.9A	150°C	2.8S min @ -3.9A	-2 to -4V	1.7 max
		IRF9640	IR	-11A	-200V	125W	0.5R max @ -6.6A	150°C	4.1S min @ -6.6A	-2 to -4V	1 max

Triacs

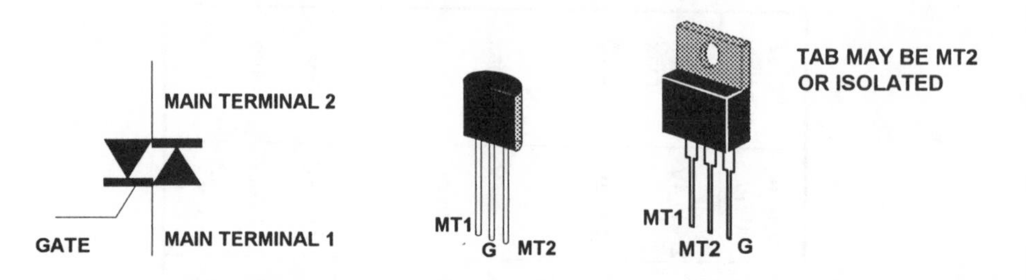

The triac is a semiconductor switch device which can control ac power. It is effectively two thyristors back to back with a common gate terminal. Before it is triggered, the triac will block current in either direction. It may be triggered into conduction by positive or negative gate current and current may then flow in either direction between main terminals 1 and 2. Once the device is triggered, the gate signal may be removed and conduction will continue until the main terminal current falls below the holding current at which point the device reverts to the blocking condition. There are four possible operating modes (triggering is always with respect to MT1).

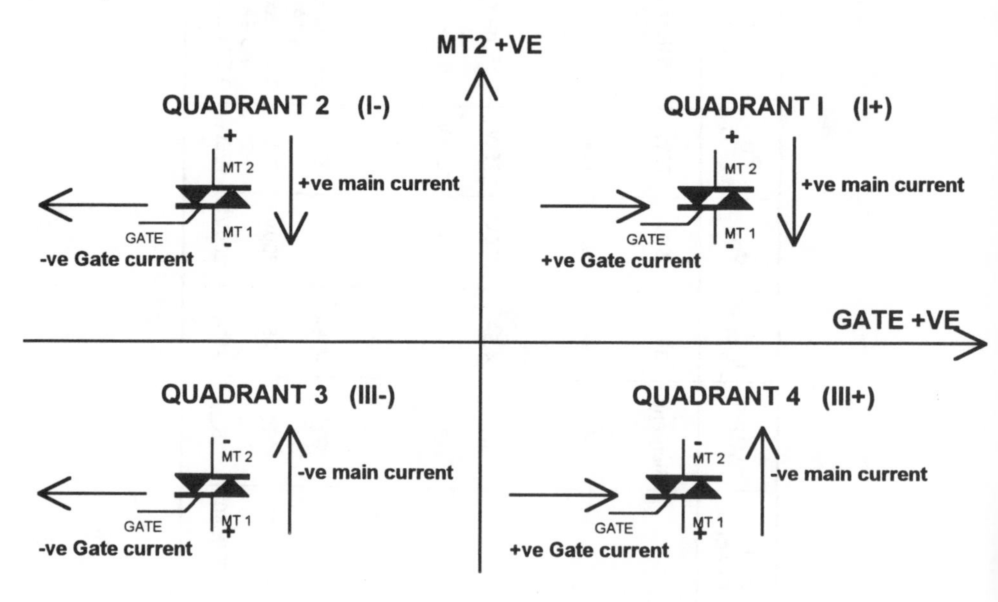

Rules of thumb

- The gate current is always derived with respect to MT1 (this may be thought of as the cathode).
- Quadrant 4 is usually less sensitive to triggering than the other three.
- When gate current is derived from MT2, the triac is operating in quadrants 1 and 3 only.
- Always use a gate to MT1 resistor to assist turn-off and reduce false triggering (1k).

Uses of a triac

Electronic switch

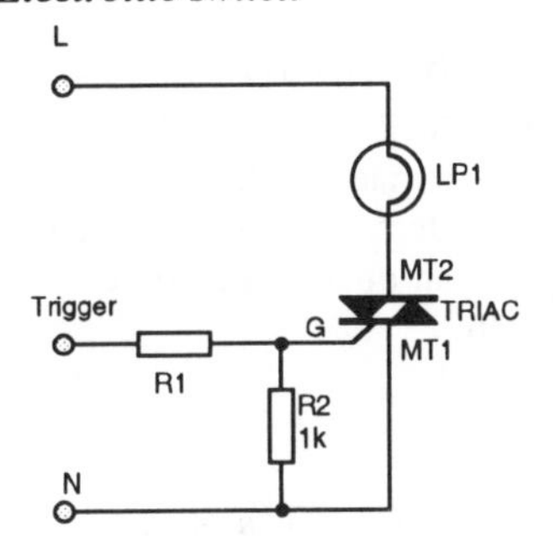

The triac may be used to switch a load such as an incandescent lamp on an ac supply. The triac is always triggered with respect to the main terminal 1 and in the circuit opposite the trigger signal may be a positive or a negative dc voltage or an ac voltage.

Lamp dimmer

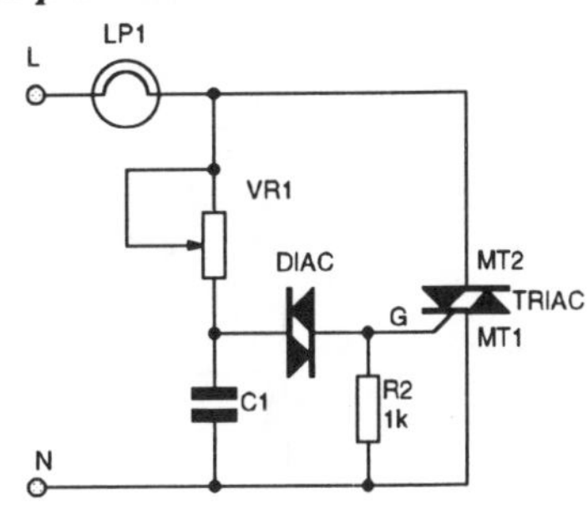

The triggering signal for the triac is derived from its main terminal 2 (MT2) VR1 and C1 provide a delay in triggering.

Triac voltage rating letter codes

Some manufacturers such as Philips signify the voltage rating of a triac device by a numbered suffix actually detailing the maximum voltage e.g. BT136-600 is a 600V device. Many manufacturers use the following suffix letter code convention:

SUFFIX LETTER	VOLTAGE RATING	SUFFIX LETTER	VOLTAGE RATING
Y	30V	M	600V
F	60V	S	700V
A	100V	N	800V
B	200V	P	1000V
C	300V	V	1200V
D	400V	Z	1400V

Testing a triac

Testing a triac is a similar process to testing a transistor (see p. 76). The MT1 to gate test gives a reading similar to a forward biased P-N junction but a greater deflection and the same in either direction. A test should be conducted with a known good triac so that the normal deflection may be recorded. The MT1-MT2 test should show no deflection in either direction, however, there will be no difference between a normal reading and an open circuit triac. The final test must be made with the device connected into the equipment.

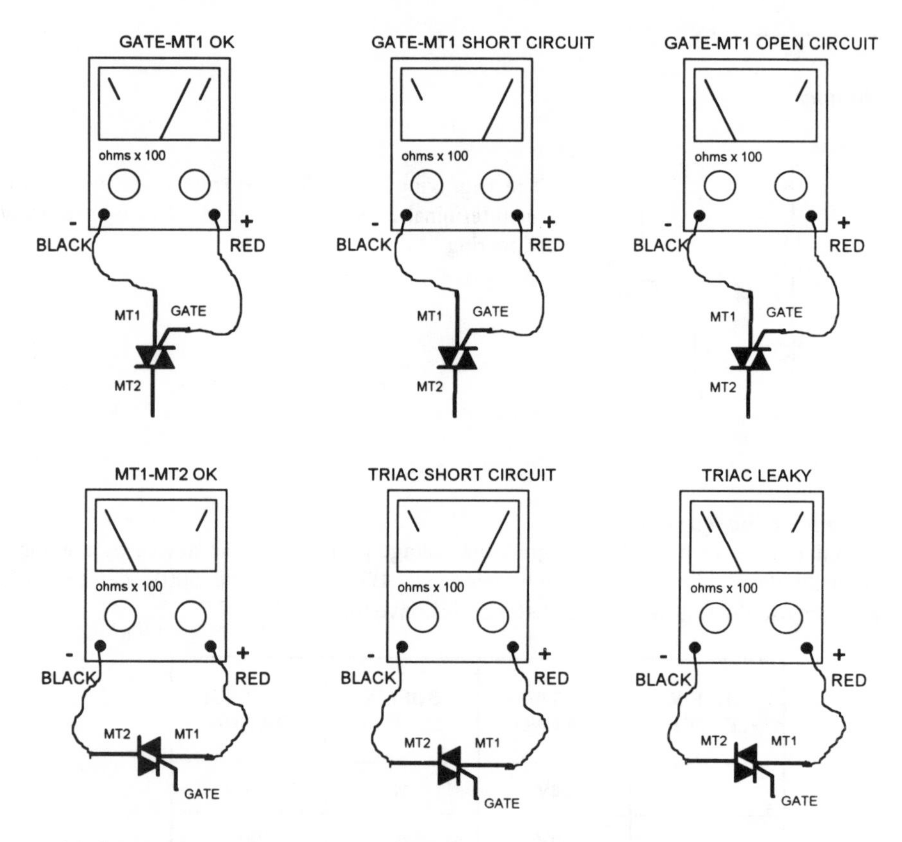

Ensure that fingers do not touch any of the leads.

Common TO220 isolated triacs

OUTLINE	TYPE No	MANFR	I_T (RMS)	I_{TSM}	V_{DRM}	dV/dt	dV/dt_c	V_{GT}	I^2t	di/dt	I_{GT} MT2+ G+	I_{GT} MT2+ G -	I_{GT} MT2- G -	I_{GT} MT2- G+	Tj °C	°C /W Junc-case
TO220 tab is isolated	BT136F-600	Philips	4A	25A	600V	100V/us	10V/us	1.5V	$4A^2s$	10A/us	35mA	35mA	35mA	70mA	110	-
	T0409MJ	TAG	4A	35A	600V	50v/us	2V/us	2.5V	$6.1A^2s$	-	10mA	10mA	10mA	10mA	125	5.75
	BTA06-600BW	Thomson	6A	60A	600V	500V/us	-	1.5V	$18A^2s$	20A/us	50mA	50mA	50mA	-	125	4.3
	T0609MJ	TAG	6A	55A	600V	50V/us	2V/us	2.5V	$15A^2s$	-	10mA	10mA	10mA	10mA	125	4.75
	BTA08-600BW	Thomson	8A	80A	600V	500V/us	-	1.5V	$32A^2s$	20A/us	50mA	50mA	50mA	-	125	4.3
	T0810MJ	TAG	8A	100A	600V	200V/us	2V/us	2.5V	$50A^2s$	-	25mA	25mA	25mA	25mA	125	3.4
	BT137F-600	Philips	8A	55A	600V	100V/us	10V/us	1.5V	$15A^2s$	20A/us	35mA	35mA	35mA	70mA	110	-
MT1 MT2 G	T1010MJ	TAG	10A	110A	600V	200V/us	2V/us	2.5V	$60A^2s$	-	25mA	25mA	25mA	25mA	125	3.4
	BT138F-600	Philips	12A	90A	600V	100V/us	10V/us	1.5V	$40A^2s$	30A/us	35mA	35mA	35mA	70mA	110	5.5
	BTA12-600BW	Thomson	12A	120A	600V	500V/us	-	1.5V	$72A^2s$	20A/us	50mA	50mA	50mA	-	125	3.3
Philips devices are a SOT186 package	T1212MJ	TAG	12A	120A	600V	500V/us	5V/us	2.5V	$72A^2s$	-	50mA	50mA	50mA	50m	125	2.8

In general: Suffix B=200V, D=400V, M=600V, N=800V

Source:- Philips, TAG, SGS-Thomson

Common TO220 non-isolated triacs

OUTLINE	TYPE No	MANFR	I_T (RMS)	I_{TSM}	V_{DRM}	dv/dt	dv/dt c	V_{GT}	I_{GT} MT2+ G+	MT2+ G -	MT2- G -	MT2- G+	Tj °C	°C /W Junc-case
TO220	BT136-600	Philips	4A	25A	600V	100V/us	10V/us	1.5V	35mA	35mA	35mA	70mA	125°C	3.7°C/W
tab is MT2	TIC206D	Texas	4A	30A	400V	-	-	-						
	T0505MH	TAG	5A	40A	600V	10V/us	1V/us	2.5V	5mA	5mA	5mA	5mA		4°C/W
	BTB06-400BW	Thomson	6A	60A	400V	-	-	-	50mA	50mA	50mA	50mA		-
	TIC216M	Texas	6A	80A	600V	-	-	-						-
	BT137-600	Philips	8A	55A	600V	100V/us	10V/us	1.5V	35mA	35mA	35mA	70mA		2.4°C/W
	T0805MH	TAG	8A	70A	600V	10V/us	1V/us	2.5V	5mA	5mA	5mA	5mA		3°C/W
	TIC225M	Texas	8A	80A	600V	-	-	2V	5mA	20mA	10mA	30mA		-
MT1	TIC226M	Texas	8A	80A	600V	-	-	2V	50mA	50mA	50mA	20mA		-
MT2 G	BTB08-600BW	Thomson	8A	80A	600V	500V/us	7V/us	1.5V	50mA	50mA	50mA	-		3.5°C/W
	T1010MH	TAG	10A	100A	600V	200V/us	2V/us	2.5	50mA	50mA	50mA	50mA		2.5°C/W
	BT138-600	Philips	12A	90A	600V	100V/us	10V/us	1.5V	35mA	35mA	35mA	70mA		2°C/W
	TIC236M	Texas	12A	100A	600V	-	-	2V	50mA	50mA	50mA	-		-
	BTB12-600BW	Thomson	12A	120A	600V	500V/us	12V/us	1.5V	50mA	50mA	50mA	-		2.7°C/W

In general: Suffix B = 200V, D = 400V, M = 600V, N = 800V

Source:- Philips, TAG, Texas, SGS-Thomson

Operational amplifiers

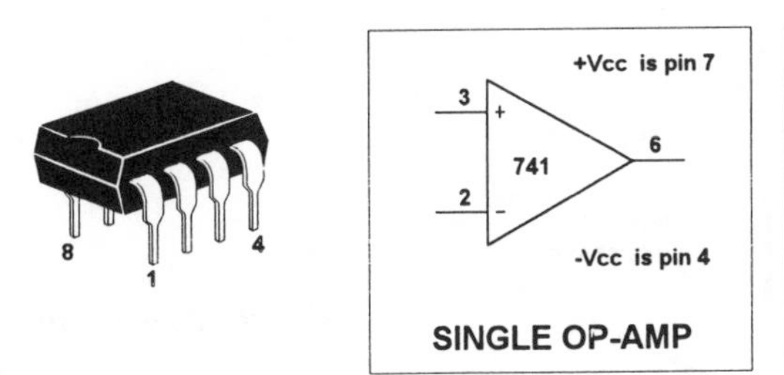

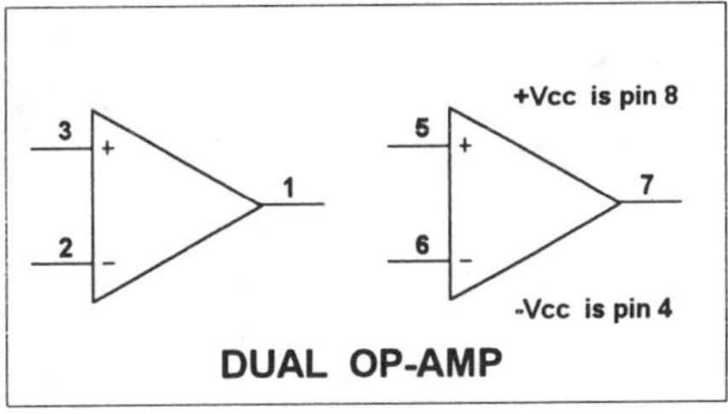

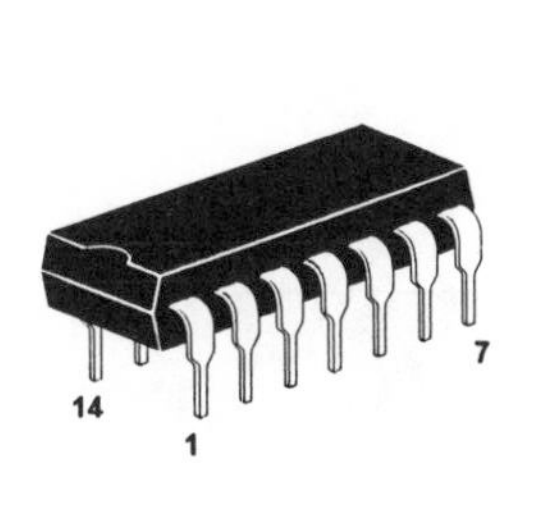

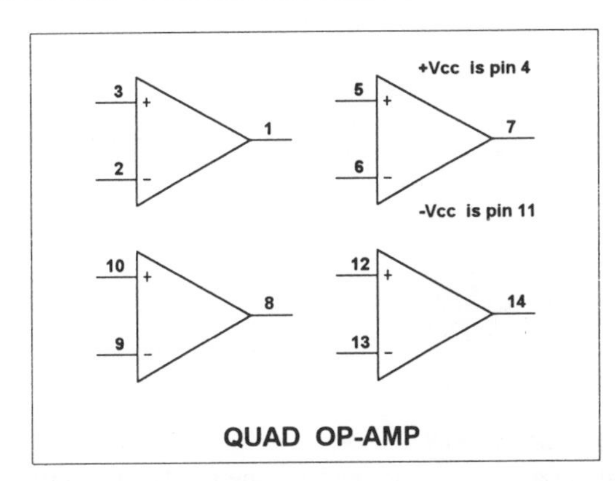

An operational amplifier (op-amp) is a high-gain differential amplifier which is capable of amplifying from dc up to frequencies in the order of 100kHz to 1 MHz depending upon the type used. It has a single-ended output i.e. the output is a voltage with respect to 0V of the power supply. There are two inputs however, one marked + and one marked - , and the op-amp is designed to amplify the voltage difference between them. When the + input is made more positive than the - input, the output goes positive with respect to 0V.

Most op-amps require positive and negative power rails typically +15V and -15V dc.

Rules of thumb

- The op-amp has very high gain, very high input impedance, and very low output impedance.
- The output gives a voltage (with respect to 0V) which is proportional to the difference in voltage between the two inputs; when the positive input is more positive than the negative input, the output voltage will be positive with respect to 0V.
- A dual rail op-amp can only operate correctly if the input signals are within approximately 3V of the power supply rails.
- A dual rail op-amp can only give an output voltage to within approximately 3V of the power supply rails.
- The impedance levels at the two inputs should be made the same to reduce input offset current effects.

Op-amp parameters

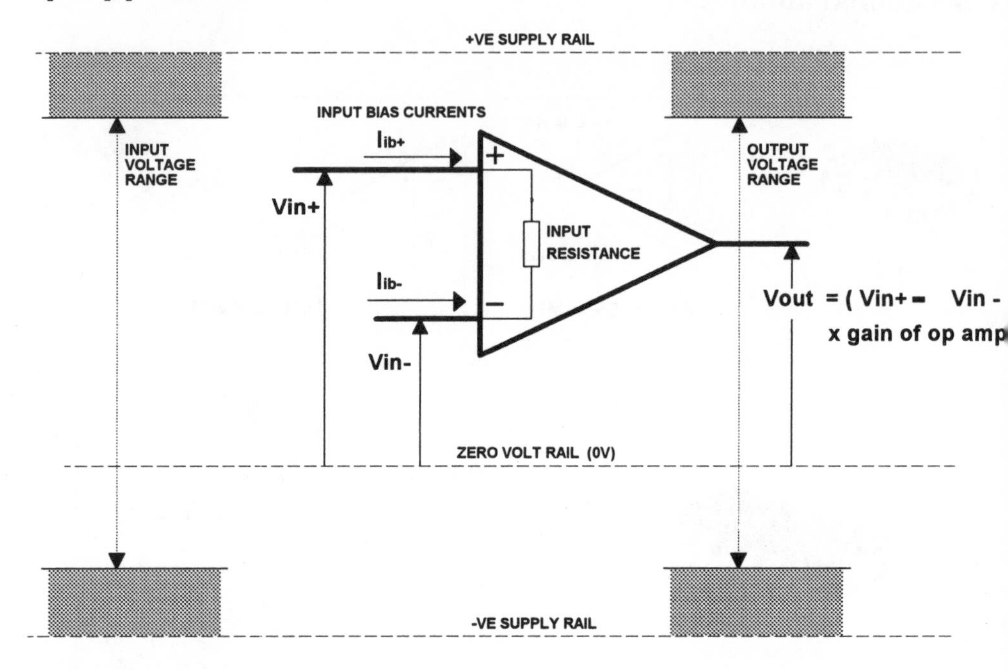

Supply voltage rails

Most op-amps are designed to operate from positive and negative dc supply rails. Typically these are +15V and -15V. The op-amp parameters will be specified at the nominal power supply rails. Most manufacturers specify absolute maximum supply rails of typically 3V above the nominal.

Supply quiescent current

This is the total current an op-amp will draw from both supply rails (at their nominal value) when no load is connected to the output.

Input offset voltage (Vos)

If the difference between the two inputs is made equal to zero, in practice the output voltage will not be zero due to slight manufacturing differences in the transistors. A very small voltage difference must be applied to the inputs to make the output zero. This input voltage is termed the input offset voltage.

Input offset voltage drift

The above parameter will drift with changes in temperature. Manufacturers quote a typical figure in microvolts per ^{o}C.

Common -mode input voltage range (CM range)

In most applications of op-amps, the two input voltages are almost the same, however they must not be allowed to fall outside of the common mode input voltage range otherwise the op-amp may not operate correctly or at worst be permanently damaged.

Differential mode input voltage range (DM range)

In most op-amp applications, the two inputs are within a few millivolts of each other. Some applications such as comparators may call for the two inputs to be different by several volts. The design of the op-amp may not allow this because of input protection diodes or the design of the input transistor stages. The DM range specifies the maximum allowable voltage difference between the two inputs.

Common mode rejection ratio CMRR

This ratio specifies how good an op-amp is at rejecting (not amplifying) signals that are present at both inputs simultaneously. The CMRR is the ratio of the common mode input voltage to the output voltage which is due to the common mode voltage. It is specified in decibels (20Log V ratio).

Input bias current

The input transistors of the op-amp require small constant currents so that they can operate correctly and these currents usually flow into the amplifier input terminals. They are termed input bias currents $\mathbf{I_{ib+}}$ and $\mathbf{I_{ib-}}$ and are specified with the output voltage = 0. Ideally these currents should be equal, however, in practice they are different and the difference is termed input offset current. Data sheets specify the average of the two input bias currents.

Input offset current

The difference between $\mathbf{I_{ib+}}$ and $\mathbf{I_{ib-}}$ when the output voltage is zero.

Input resistance

This is the resistance measured at one input when the other input is connected to 0V.

Slew rate

This is the maximum rate of change of voltage the output can achieve when the op-amp has feedback connected and a step change in input voltage is applied. The slew rate limits the sine-wave frequency at which an op-amp can amplify and give the highest peak-to-peak output voltage without distortion. Slew rate is specified in volts per microsecond. An amplifier which is showing slew rate limitations is seen to give a triangular-wave output for a sine-wave input.

Maximum peak output voltage range

The output of an op-amp can only operate within a certain voltage range without saturating to one of the supply rails or becoming non-linear. The range is specified as a positive and negative voltage limit and is dependent upon the load resistance connected to the output of the op-amp.

Differential voltage gain

This is the ratio of the output voltage to the differential input voltage when no feedback is connected around the op-amp. It is typically in the region of 200,000.

Load

This is the resistance connected between the output and 0V which is used when specifying the output voltage range.

Uses of an op-amp

Inverting amplifier

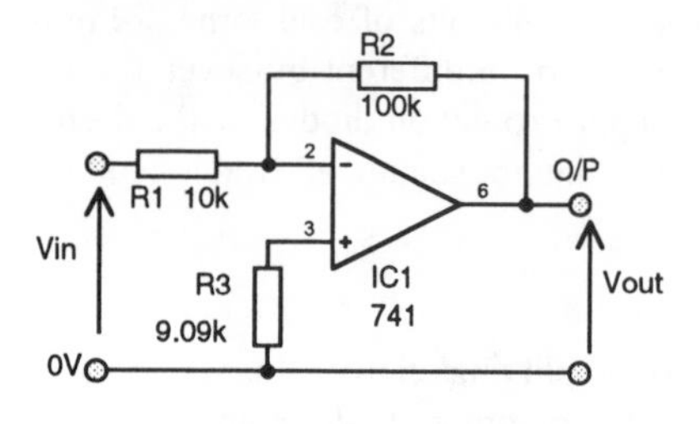

The voltage gain is equal to:

$$\text{GAIN} = \frac{\text{Vout}}{\text{Vin}} = -\frac{\text{R2}}{\text{R1}}$$

R3 should be = R1 x R2 /(R1 + R2) so as to minimize effects of input offset current.

Non-inverting amplifier

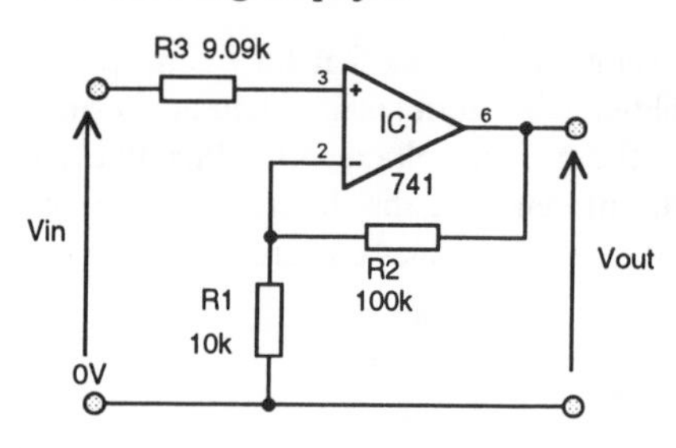

The voltage gain is equal to:

$$\text{GAIN} = \frac{\text{Vout}}{\text{Vin}} = 1 + \frac{\text{R2}}{\text{R1}}$$

R3 should be = R1 x R2 /(R1 + R2) so as to minimize effects of input offset current.

Voltage follower

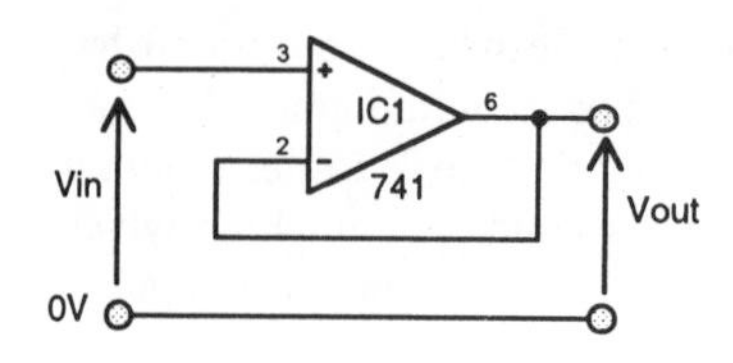

The output voltage equals the input voltage without inversion. The main purpose of the voltage follower is to introduce current gain to buffer a load such as a meter from another circuit which may have a high enough voltage output signal but not enough current drive capability. The voltage follower has a very high input impedance and very low output impedance.

Summing amplifier

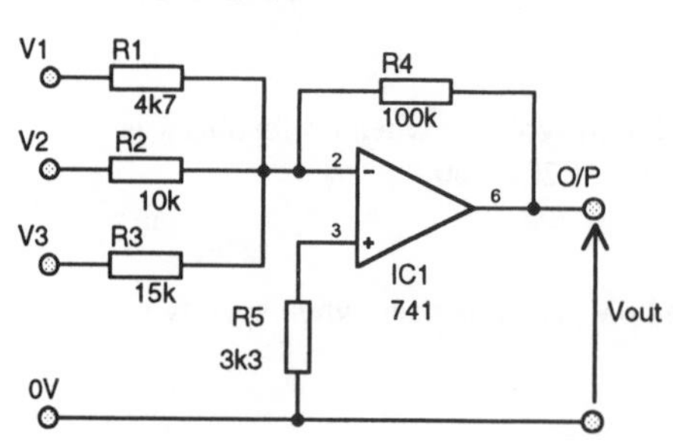

The circuit opposite adds all three input voltages V1, V2 and V3 according to the equation:

$$\text{Vout} = -\text{R4}\left(\frac{\text{V1}}{\text{R1}} + \frac{\text{V2}}{\text{R2}} + \frac{\text{V3}}{\text{R3}}\right)$$

In order to minimize errors due to input offset current effects: 1/R5 = 1/R1 + 1/R2 + 1/R3. The circuit may be used as an audio signal mixer.

Differential amplifier

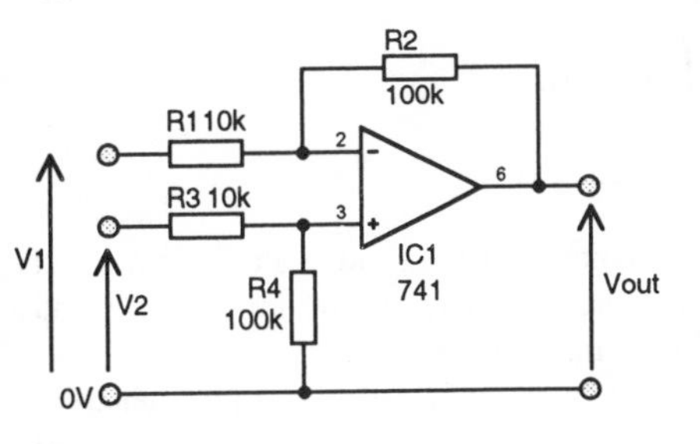

The differential amplifier (or subtractor) gives an output voltage which is proportional to the difference between the two inputs V1 and V2. In the example opposite, R2 = R4 and R1 = R3 and:

$$\mathbf{Vout} = \frac{\mathbf{R2}}{\mathbf{R1}}\left(\mathbf{V2 - V1}\right)$$

Differentiator

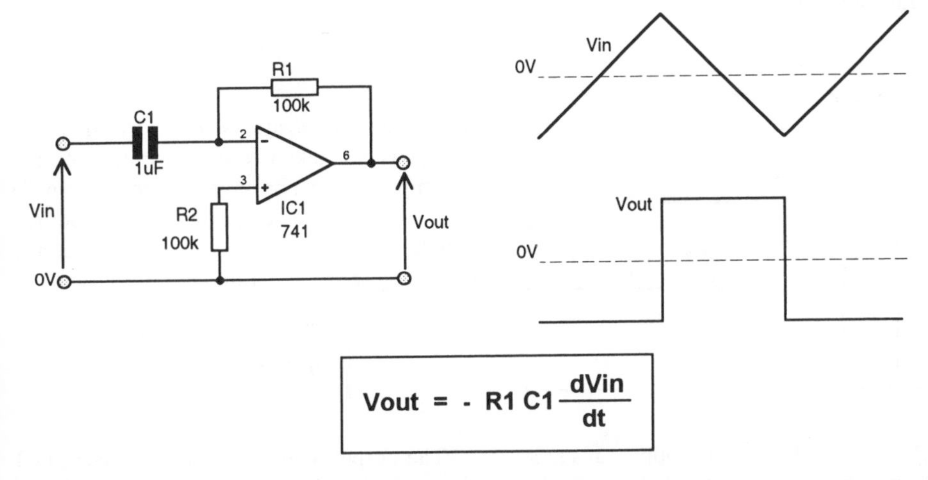

$$\mathbf{Vout = - R1\,C1 \frac{dVin}{dt}}$$

The differentiator (not to be confused with the differential amplifier) gives an output voltage which is proportional to the rate of change of the input voltage Vin. In practice, the above circuit suffers from the effects of high frequency noise and instability. In order to combat these effects a small resistor Rs is placed in series with C1. The integrator is then only applicable to frequencies much less than 1/(2pC1Rs).

Integrator

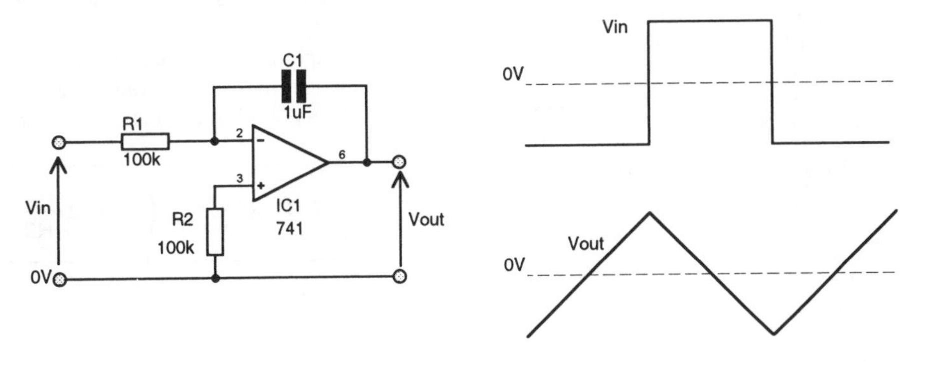

$$V_{out} = -\frac{1}{R1\,C1}\int V_{in}\,dt$$

The output voltage is the negative integral of the input voltage; if the input voltage is a steady dc negative value, the output will ramp positive in a linear manner. If the input is a steady positive dc value, the output will ramp negative in a linear manner.

In practice the above circuit suffers from the effects of small input offset voltages which are eventually integrated up and cause the output of the op amp to saturate to one of the supply rails. This problem is overcome by connecting a large value resistor (Rp) in parallel with C1 to define the gain at dc.

Voltage to constant current converter

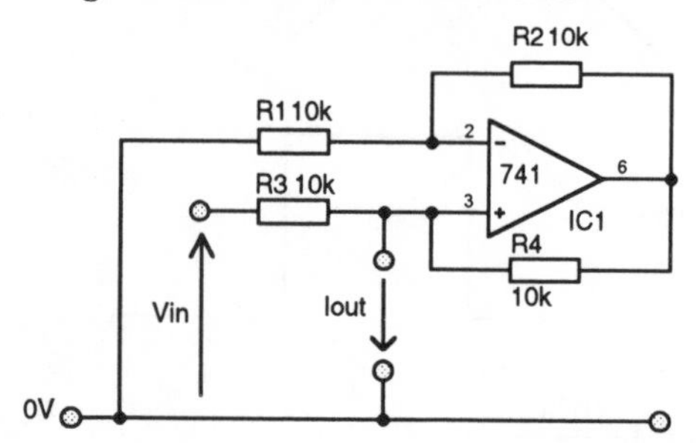

The circuit opposite is novel in that the output is not taken from the output of the op-amp but the non-inverting input (pin 3) The output is a constant current source which is proportional to the input voltage according to the equation:

$$I_{out} = \frac{V_{in}}{R3}$$

R1 = R2 = R3 = R4 and the supply rails are ±15V. The output load connected between pin 3 and 0V may be any resistance between a short circuit and 5k. Vin may be between -10V and +10V. The circuit above gives a constant output current of +1mA when Vin = +10V.

Astable multivibrator (square wave oscillator)

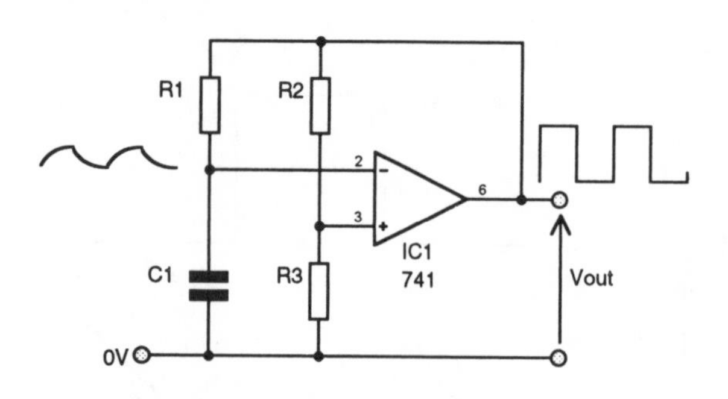

The circuit opposite produces an equal mark-space ratio square wave output with a peak to peak amplitude of approximately 25V when the supply rails are ±15V. The period of oscillation is given as:

$$2\,C1\,R1\,\text{Log}_e\left(1 + 2\frac{R3}{R2}\right)$$

Gyrator

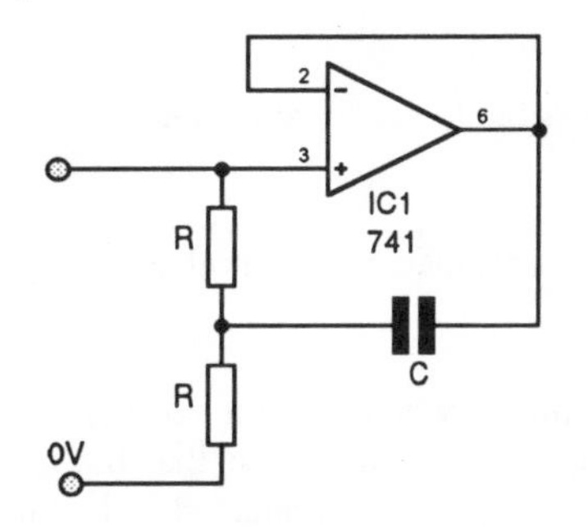

The gyrator synthesizes an inductor by the use of negative feedback via a capacitor. The input impedance looking into the two input terminals is:

$$2R + j\omega CR^2$$

i.e. a resistance of 2R in series with an inductance equal to CR^2.

Sine-wave oscillator

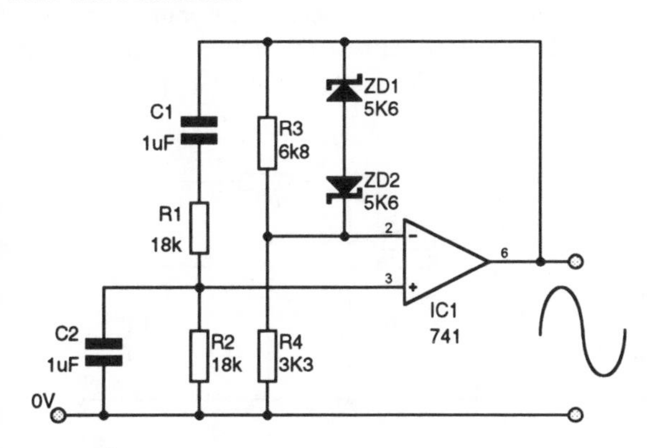

R1, R2, C1 and C2 form a Wien network which sets the frequency of oscillation at:

$$f_{out} = \frac{1}{2\pi C1R1}$$

where R1 = R2 and C1 = C2

At the oscillation frequency, the voltage fed back to pin 3 of the op-amp is in phase with the output and attenuated to 1/3.

Precision half-wave rectifier

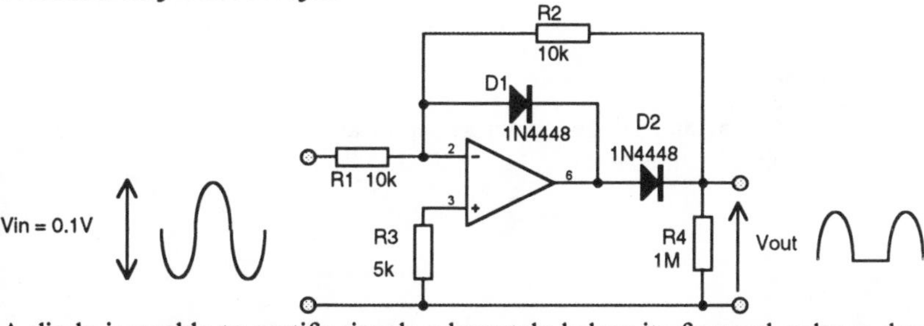

A diode is unable to rectify signals adequately below its forward voltage drop of about 0.6V and ac signals of an amplitude of a few volts are distorted accordingly. The circuit above is able to half-wave rectify signals down to a fraction of a volt without distortion but gives 180 degree inversion. When the input signal is negative, pin 6 of the op-amp goes positive and causes D2 to conduct via R2 and R1. Vout is the same amplitude as Vin only inverted. The op-amp is always more positive than Vout by the forward voltage drop of D2. The non-linearity of D2 is compensated for by the op-amp as D2 is inside the feedback loop. When the input signal Vin goes positive, the output of the op-amp goes negative and D1 conducts. This causes D2 to be reverse biased and thus Vout = 0V (the same voltage as pin 2 of the op amp i.e. virtual earth). R4 gives a path for leakage current of D2.

Design pitfalls when using op-amps

- Failure to de-couple the power supply rails as close as possible to the op-amp. This may cause the op-amp to oscillate or malfunction.
- Failure to balance the impedance levels at the two inputs. This may cause an offset voltage due to the input bias currents.
- 0V rail problems; the two inputs are connected to 0V points which have a voltage drop between them. In the circuit on the left, the power supply is connected so that the load current through R4 has to flow along the wire to which the inputs are connected. The load current produces a voltage drop DV across the small resistance of the wire and this appears in series with the signal V1. The op amp amplifies DV as well as the signal and the op amp may saturate, oscillate or latch up. The problem is solved by the circuit layout on the right where the power supply is connected so that the load current through R4 returns to the supply without flowing through the wire connected to the inputs of the op-amp.

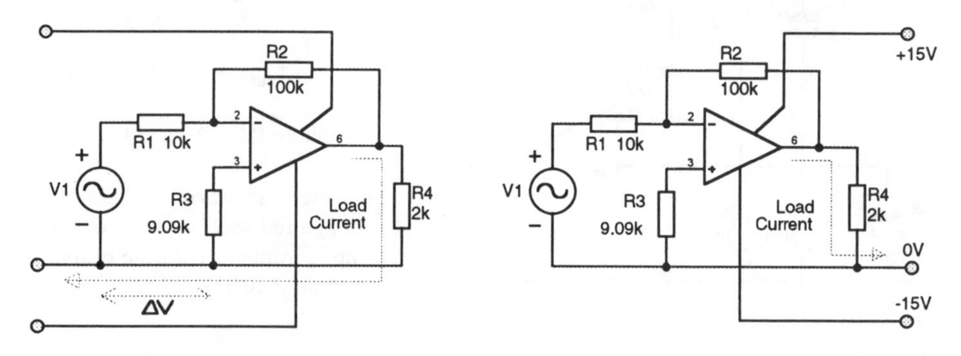

Testing an op-amp

An op-amp may not be tested using an analogue meter as per a transistor; the acid test is to connect the op-amp in a circuit it is suitable for.

Single op-amps

All parameters are at 25°C and are worst case unless detailed otherwise

	SUPPLY		INPUT VOLTAGE					INPUT CURRENT				OUTPUT VOLTAGE			
TYPE No	V_{RAIL}	I_Q	V_{OS}	V_{OS} DRIFT typ	CM RANGE	DM RANGE	CMRR	BIAS	I_{OS}	I_{OS} DRIFT	RES	SLEW RATE	RANGE	GAIN	LOAD
OP07C	±15V	5mA	0.15mV	1.8uV/ °C	±13Vmax	±30Vmax	100dB	±7nA	6nA	50pA/ °C	8M	0.1V/uS	±11.5V	100V/mV	2k
OP27C	±15V	5.6mA	0.1mV	1.8uV/ °C	±11Vmax	-	100dB	±80nA	75nA	10pA/ °C	2G	1.7V/uS	±11.5V	700V/mV	2k
OP37C	±15V	5.6mA	0.1mV	1.8uV/ °C	±11Vmax	-	100dB	±80nA	75nA	10pA/ °C	2G	11V/uS	±11.5V	700V/mV	2k
TL061C	±15V	.25mA	15mV	10uV/ °C	±11Vmax	±30Vmax	70dB	400pA	200pA	-	1T	1.5V/uS	±10V	3V/mV	10k
TL071C	±15V	2.5mA	10mV	10uV/ °C	±11Vmax	±30Vmax	70dB	200pA	100pA	-	1T	8V/uS	±10V	25V/mV	2k
TL081C	±15V	2.8mA	15mV	10uV/ °C	±11Vmax	±30Vmax	70dB	400pA	200pA	-	1T	8V/uS	±10V	15V/mV	2k
LM301A	±15V	3mA	7.5mV	30uV/ °C	±12Vmax	±30Vmax	70dB	250nA	50nA	1nA/ °C	0.5M	0.4V/uS	±10V	25V/mV	2k
LM308A	±15V	0.8mA	7.5mV	30uV/ °C	±14Vmax	-	80dB	7nA	1nA	10pA/ °C	10M	0.3V/uS	±13V	25V/mV	10k
AD548C	±15V	0.2mA	0.25mV	2uV/ °C	±11Vmax	±20Vmax	76dB	10pA	5pA	-	1T	1.8V/uS	±11V	150V/mV	5k
uA709M	±15V	5.5mA	5mV	3uV/ °C	±8Vmax	±5Vmax	70dB	500nA	200nA	-	0.15M	-	±10V	25V/mV	2k
uA741C	±15V	2.8mA	6mV	10uV/ °C	±12Vmax	±30Vmax	70dB	500nA	200nA	0.2nA/ °C	0.3M	0.5V/uS	±10V	20V/mV	2k
CA3130E	±8V	2mA	8mV	-	-	±8V	90dB	5pA	0.5pA	-	1.5T	10V/uS	±10V	300V/mV	2k
CA3140E	±15V	4mA	5mV	-	-	±8V	90dB	10pA	0.5pA	-	1.5T	9V/uS	±10V	100V/mV	2k
NE5534	±15V	6.5mA	2mV	5uV/ °C	±12Vmax	±0.5Vmax	70dB	500nA	200nA	0.2nA/ °C	50k	13V/uS typ	±12V	50V/mV	600

Voltage regulators

100mA positive voltage regulators

PIN-OUT	TYPE No	INPUT VOLTAGE RANGE	OUTPUT VOLTAGE 1mA<Iout<70mA min<Vin<max	LINE REGULATION min<Vin<max	LOAD REGULATION 1mA<Iout<100mA	θJA	CIRCUIT
TO92 pin view input output common	78L05AC 78L08AC 78L12AC 78L15AC 78L18AC 78L24AC	7V to 20V 10.5V to 23V 14.5V to 27V 17.5V to 30V 20.7V to 33V 27V to 38V	4.75Vmin, 5.25Vmax 7.6Vmin, 8.4Vmax 11.4Vmin, 12.6V max 14.25Vmin,15.75Vmax 17.1Vmin, 18.9Vmax 22.8Vmin, 25.2Vmax	75mV max 100mV max 180mV max 250mV max 275mV max 300mV max	60mV max 80mV max 100mV max 150mV max 170mV max 200mV max	180 ^{o}C/W	Input 78LxxAC Output 330nF 100nF Common

100mA negative voltage regulators

PIN-OUT	TYPE No	INPUT VOLTAGE RANGE	OUTPUT VOLTAGE 1mA<Iout<70mA min<Vin<max	LINE REGULATION min<Vin<max	LOAD REGULATION 1mA<Iout<100mA	θJA	CIRCUIT
TO92 pin view output common input	79L05AC 79L12AC 79L15AC 79L18AC 79L24AC	-7V to -20V -14.5V to -27V -17.5V to -30V -20.7V to -33V -27V to -38V	-4.75Vmin, -5.25Vmax -11.4Vmin, -12.6V max -14.25Vmin,-15.75Vmax -17.1Vmin, -18.9Vmax -22.8Vmin, -25.2Vmax	60mV max 45mV max 45mV max 50mV max 60mV max	50mV max 100mV max 125mV max 150mV max 200mV max	180 ^{o}C/W	Input 79LxxAC Output 330nF 100nF Common

Source - National Semiconductor

1A positive voltage regulators

PIN-OUT TOP VIEW	TYPE No	INPUT VOLTAGE RANGE	OUTPUT VOLTAGE 5mA<Iout<1A min<Vin<max	LINE REGULATION min<Vin<max	LOAD REGULATION 5mA<Iout<1.5A	θJA	θJC	CIRCUIT
TO220 tab is common; input, common, output	7805	8V to 20V	4.65Vmin, 5.35Vmax	50mV max	100mV max	65 °C/W	5 °C/W	Input, 78xx, Output, 330nF, Common
	7808	11.5V to 23V	7.6Vmin, 8.4Vmax	80mV max	100mV max			
	7812	15.5V to 27V	11.4Vmin, 12.6V max	120mV max	120mV max			
	7815	18.5V to 30V	14.25Vmin,15.75Vmax	150mV max	150mV max			
	7818	22V to 33V	17.1Vmin, 18.9Vmax	180mV max	180mV max			
	7824	28V to 38V	22.8Vmin, 25.2Vmax	240mV max	240mV max			

1A negative voltage regulators

PIN-OUT TOP VIEW	TYPE No	INPUT VOLTAGE RANGE	OUTPUT VOLTAGE 5mA<Iout<1A min<Vin<max	LINE REGULATION min<Vin<max	LOAD REGULATION 5mA<Iout<1.5A	θJA	θJC	CIRCUIT
TO220 tab is input; common, input, output	7905C	-7V to -20V	-4.75Vmin, -5.25Vmax	100mV max	100mV max	65 °C/W	5 °C/W	Input, 78xx, Output, 330nF, Common
	7908C	-10.5V to -23V	-7.6Vmin, -8.4Vmax	160mV max	160mV max			
	7912C	-14.5V to -27V	-11.4Vmin, -12.6V max	240mV max	240mV max			
	7915C	-17.5V to -30V	-14.25Vmin,-15.75Vmax	300mV max	300mV max			
	7918C	-21V to -33V	-17.1Vmin, -18.9Vmax	360mV max	360mV max			
	7924C	-27V to -38V	-22.8Vmin, -25.2Vmax	480mV max	480mV max			

555 timer

Equivalent circuit

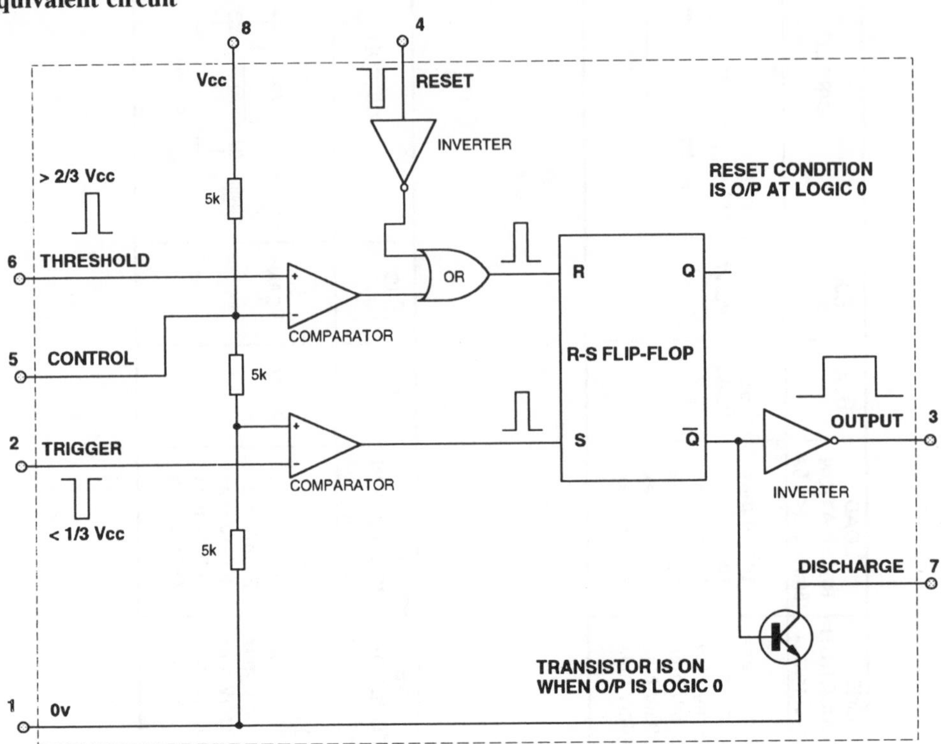

NE555

Power supply: 4.5V to 16V (always fit a de-coupling capacitor directly across IC)
Supply current: 15mA max at Vcc = 15V, 6mA max at Vcc = 5V
Output drive current: ± 200 mA

Monostable mode

+15V
RT
TRIGGER I/P
555
IC1
o/p
1.1RtCt
CT
C1
0.1uF
0V

Astable mode

+15V
R1
R1>1k
R2
555
IC2
o/p
0.693 (R1 + R2) CT
0.693 R2 CT
CT
C1
0.1uF
0V

Logic circuits

Logic circuits are digital computational circuits based on the functions: AND, OR, NAND, NOR, and memory and counting devices which may be connected to form a digital control or computational system such as a industrial production line controller or a computer. One two logic states are used, 1 or 0. Logic circuits were first implemented using thermionic valves and then later, using discrete transistor circuits. The first integrated circuit "family" of logic circuits was transistor-transistor logic or TTL which was based on multi-emitter bipolar integrated transistors. The table below summarizes the electrical parameters of the logic families currently available today.

	TTL				CMOS			ECL	
FAMILY	**74**	**74LS**	**74AS**	**74ALS**	**4000B**	**74HC**	**74HCT**	**10k**	**100k**
SUPPLY	5.0V	5.0V	5.0V	5.0V	3-15V	5.0V	5.0V	-5.2V	-4.2V
V_{IH} (min)	2.0V	2.0V	2.0V	2.0V	$0.7V_{CC}$	3.5V	2.0V	-1.2V	-1.2V
V_{IL} (max)	0.8V	0.8V	0.8V	0.8V	$0.3V_{CC}$	5.0V	5.0V	-5.2V	-4.2V
V_{OH} (min)	2.4V	2.7V	2.7V	2.7V	V_{CC}-0.05	4.9V	4.9V	-0.9V	-0.9V
V_{OL} (max)	0.4V	0.5V	0.5V	0.5V	0.05V	0.1V	0.1V	-1.7V	-1.7V
I_{IH} (max)	0.04mA	0.02mA	0.2mA	0.02mA	0.3uA	0.1uA	0.1uA		
I_{IL} (max)	1.6mA	0.4mA	2mA	0.1mA	0.3uA	0.1uA	0.1uA		
I_{OH} (max)	0.4mA	0.4mA	2mA	0.4mA	3mA	4mA	4mA	50mA	55mA
I_{OL} (max)	16mA	8mA	20mA	8mA	3mA	4mA	4mA	50mA	55mA
mW/GATE	10mW	2mW	8mW	1mW	≅0	≅0	≅0	25mW	40mW
DELAY	10nS	10nS	1.5nS	4nS	60nS	30nS	10nS	2nS	0.75nS
FAN-OUT	10	20	10	20	>100	>100	>100	10	10

All parameters are at 25 ^{O}C and nominal power supply voltage (15V for CMOS 4000B).

Manufacturer's prefix letters

CD	HARRIS
DM	NATIONAL SEMICONDUCTOR
HD	HITACHI
HEF	PHILIPS COMPONENTS
M74	SGS THOMSON
MC	MOTOROLA
MM	NATIONAL SEMICONDUCTOR
N74F	PHILIPS COMPONENTS
PC74	PHILIPS COMPONENTS
SN	TEXAS INSTRUMENTS
T	SGS-THOMSON
TC	TOSHIBA

Logic symbols and definitions

FUNCTION	SYMBOL	TRUTH TABLE A	B	C	BOOLEAN	DESCRIPTION
AND	A, B, C	0 1 0 1	0 0 1 1	0 0 0 1	A . B = C	The output is logic 1 when both inputs are logic 1 otherwise the output is logic 0
OR	A, B, C	0 1 0 1	0 0 1 1	0 1 1 1	A + B = C	The output is only logic 0 when both inputs are logic 0 otherwise the output is logic 1
NAND	A, B, C	0 1 0 1	0 0 1 1	1 1 1 0	$\overline{A . B} = C$	By De Morgan's theorem the NAND gate may be thought as an OR gate for 0s at the inputs
NOR	A, B, C	0 1 0 1	0 0 1 1	1 0 0 0	$\overline{A + B} = C$	By De Morgan's theorem the NOR gate may be thought as an AND gate for 0s at the inputs
BUFFER	A, C	0 1		0 1	A = C	The output is the same logic state as the input but may have higher current drive
INVERT	A, C	0 1		1 0	$\overline{A} = C$	The output is the inverse logic state of the input
EXCLUSIVE OR	A, B, C	0 1 0 1	0 0 1 1	0 1 1 0	$A.\overline{B} + \overline{A}.B$ = C	The output is a logic 1 when the two inputs are at different logic states

FUNCTION	SYMBOL	S R	Q $\overline{Q}$		DESCRIPTION
NAND R - S LATCH	S, R, Q, $\overline{Q}$	0 0 1 0 0 1 1 1	1 1 0 1 1 0 No change	S, R, Q, Q	The two inputs are normally held at logic 1 and the latch is set or reset by logic 0 pulses
NOR R - S LATCH	S, R, $\overline{Q}$, Q	1 1 0 1 1 0 0 0	0 0 0 1 1 0 No change	S, R, $\overline{Q}$, Q	The two inputs are normally held at logic 0 and the latch is set or reset by logic 1 pulses

Decimal to binary to hexadecimal

Hexadecimal is a number system based on the number 16 and is used as a shorthand method of recording large binary numbers; i.e.the binary number 1011 1111 1010 1000 may be represented by the hex BFA8.

DECIMAL	BINARY				HEX
	$2^3 = 8_{DECIMAL}$	$2^2 = 4_{DECIMAL}$	$2^1 = 2_{DECIMAL}$	$2^0 = 1_{DECIMAL}$	
0	0	0	0	0	0
1	0	0	0	1	1
2	0	0	1	0	2
3	0	0	1	1	3
4	0	1	0	0	4
5	0	1	0	1	5
6	0	1	1	0	6
7	0	1	1	1	7
8	1	0	0	0	8
9	1	0	0	1	9
10	1	0	1	0	A
11	1	0	1	1	B
12	1	1	0	0	C
13	1	1	0	1	D
14	1	1	1	0	E
15	1	1	1	1	F

Serial data transmission

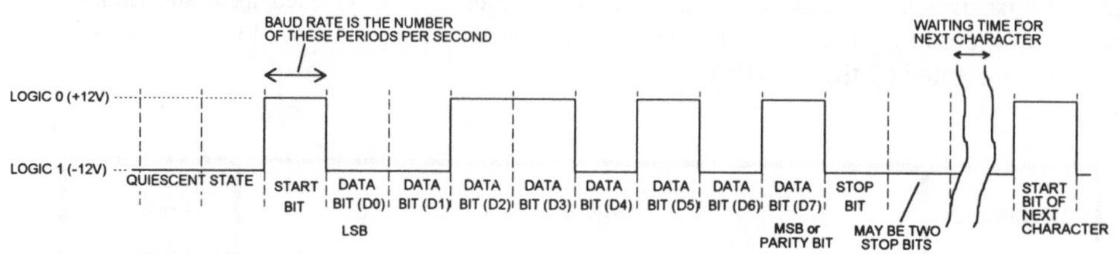

Note that logic '0' is +12V.

The above timing diagram shows the serial transmission of 1100 1010.

BAUD RATE	PERIOD OF EACH BIT
110	9.091mS
150	6.67mS
300	3.33mS
600	1.67mS
1200	833uS
2400	417uS
4800	208uS
9600	104uS
19,200	52.08uS

ASCII code

		FIRST DIGIT							
		000	001	010	011	100	101	110	111
SECOND DIGI	HEX	0	1	2	3	4	5	6	7
0000	0	NUL	DLE	SP	0	@	P	\	p
0001	1	SOH	DC1	!	1	A	Q	a	q
0010	2	STX	DC2	"	2	B	R	b	r
0011	3	ETX	DC3	#	3	C	S	c	s
0100	4	EOT	DC4	$	4	D	T	d	t
0101	5	ENQ	NAK	%	5	E	U	e	u
0110	6	ACK	SYN	&	6	F	V	f	v
0111	7	BEL	ETB	'	7	G	W	g	w
1000	8	BS	CAN	(	8	H	X	h	x
1001	9	HT	EM	)	9	I	Y	i	y
1010	A	LF	SUB	*	:	J	Z	j	z
1011	B	VT	ESC	+	;	K	[	k	{
1100	C	FF	FS	,	<	L	\	l	|
1101	D	CR	GS	-	=	M	]	m	}
1110	E	SO	RS	.	>	N	^	n	~
1111	F	SI	US	/	?	O	_	o	DEL

Example: The ASCII code for capital letter "K" would be $4B_{hex}$ or 0100 1011 binary. The first binary bit may be used as a parity check, however.

5 Sensors and transducers

Definition

A transducer (often called a sensor) is a device which responds to a signal which is in the form of a type of energy and converts it to a different form of energy or phenomenon, but retains the signal information. For example, a microphone is a transducer which converts sound energy into electrical energy. The electrical signal which the microphone produces varies in a manner which closely resembles the variations in sound pressure which it responds to. The most common types of transducer are those which convert to or from electrical energy. In general a transducer exploits some phenomenon which was discovered or invented. Transducers fall into two main categories: active and passive. In active transducers actual energy conversion occurs and the output from the transducer may be used to drive an indicating device. An example of an active transducer is a photo-electric cell. A passive transducer is only able to control the flow of energy applied to its output. An example of a passive transducer is a strain gauge. The strain gauge does not give an electrical output, but it can vary the amount of electrical energy passing through it. The table below shows some examples of transducers although it does not cover every possible type of transducer for each category of energy conversion.

ENERGY CONVERSION	TRANSDUCER	PHENOMENON
Sound to electrical	Microphone	Electromagnetic induction
Heat (temperature) to electrical	Thermocouple	Seebeck effect
Strain or displacement to electrical	Strain gauge	Variation in electrical resistance (passive)
Light to electrical	Photo-transistor	Photons cause charge carriers to be produced in a semiconductor

Rules of thumb

When choosing and using a transducer make the following considerations:

- The range of the phenomenon to be measured (e.g. the temperature range).
- The resolution or accuracy required.
- The linearity i.e. what is the relationship between the phenomenon being measured and the output signal.
- How it may be affected by phenomenon other than that being measured (e.g. the output of a strain gauge may also be affected by temperature variations).

Interfacing a transducer to electronics

Most transducers are designed to interface with an electrical or electronic circuit. The signal being sensed by a transducer may have to be displayed by a digital read-out or logged on a chart recorder. The signal may have to be sent to a control system which reacts to variations in the signal. In many cases it is not possible to connect the output of the transducer directly to an indicating or other device. The signal from the transducer will probably have to be processed in some way by an electronic circuit. The signal may be very small and require amplification, or it may have to be linearized or compensated for other unwanted effects.

Loading effects

Whenever a transducer is connected to an electronic circuit, the impedance of the circuit may "load up" the transducer and reduce the amplitude of the signal or distort its characteristic in some undesirable fashion. The most common loading effect is when the electronic circuit has too low an input resistance and simply "kills" the signal from the transducer. Some transducers such as electret microphones require an electronic circuit which has a very high input resistance.

Coupling

The output of a transducer may be varying in a very slow manner in which case it may be necessary to directly couple it to an electronic circuit which can amplify dc signals. Some transducers are only concerned with ac signals and may have to be coupled via a capacitor to remove any dc signal which may be present which could interfere with the circuit. The choice of value of the coupling capacitor depends upon the impedance of the transducer, the input impedance of circuit and the lowest frequency to be processed.

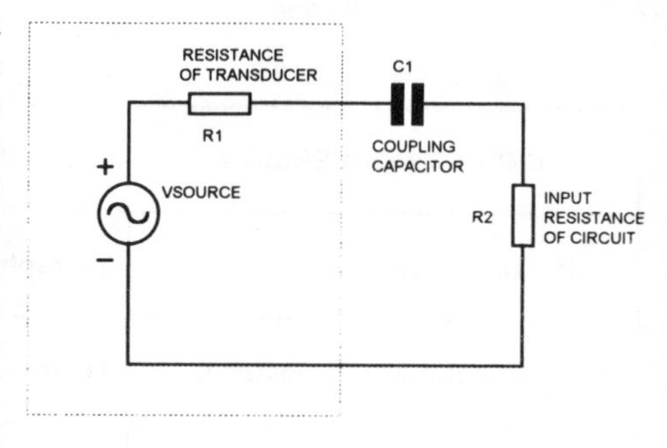

If f is the lowest frequency to be processed then the value of the coupling capacitor C1 must be chosen such that

$$R_1 + R_2 = \frac{1}{2\pi f C_1}$$

If the input resistance of the circuit R2 is more than ten times the resistance of the transducer R1, then R1 may be omitted from the equation.

Linearization

Ideally, there should be a linear relationship between the parameter being sensed (e.g. temperature) and the electrical signal produced by the transducer, otherwise a non-linear scale would be required on a meter, or the paper chart of a recorder would have to have a non-linear scale. Further processing of the electrical signal may also prove to be complicated if it is non-linear. For example, the flow of a liquid may be measured by a device called an orifice plate (sometimes called a venturi meter). As liquid flows through the venturi, a pressure difference is created. A differential pressure sensor may used to measure this. The magnitude of the flow is actually proportional to the pressure squared and so a linearization circuit is required which has a square-root characteristic.

Thermocouples

Definition
A thermocouple is two dissimilar metal wires joined at each end to form an electrical circuit. If the two junctions are at different temperatures, a voltage is produced between the wires due to the Seebeck effect.

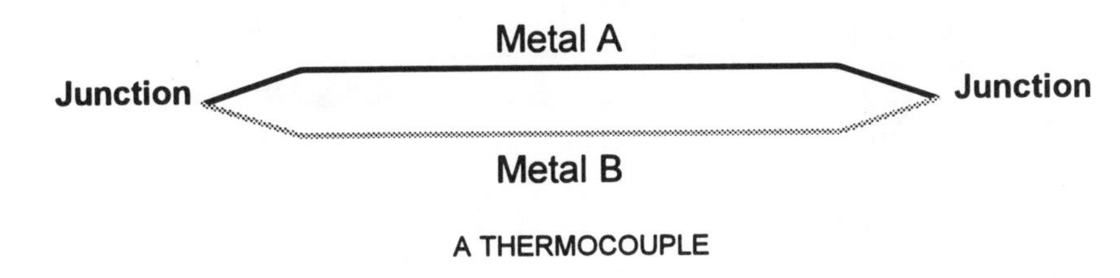

A THERMOCOUPLE

Background
The number of free electrons in a metal is dependent upon its type and its temperature. If two dissimilar metal wires are joined together at each end a small voltage exists between them. If the junctions are at the same temperature then no current will flow as the same voltage exists at each junction and one voltage exactly opposes the other. When the junctions are at different temperatures, the voltages are different and with some combinations of metals, the difference in voltages is almost proportional to the temperature difference. When an attempt is made to measure the voltage difference with a voltmeter by breaking into the circuit, other thermocouples are formed when the wires of the voltmeter are connected to the metal wires. If these new junctions are both kept at the same temperature however, the voltages produced will be equal and will cancel each other out. The voltmeter will therefore indicate the temperature difference between the two main junctions. An engineer may wish to measure the temperature of an object such as a transistor and one of the junctions of the thermocouple will be attached to the transistor. This junction is usually referred to as the **hot junction** If the other junction (referred to as the **cold junction**) is at room temperature then the voltmeter will simply indicate the temperature difference between the transistor and the room. For the voltmeter to read the absolute temperature of the transistor in degrees Celsius, it is necessary to allow for the temperature of the cold junction. This is termed **cold junction compensation.**

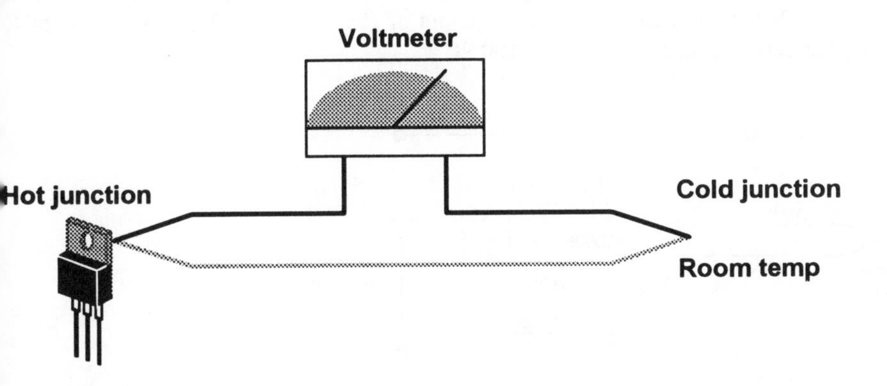

Cold junction compensation

The cold junction is maintained at 0 ^{o}C by immersing in melting ice. The hot junction will then read directly in ^{o}C.

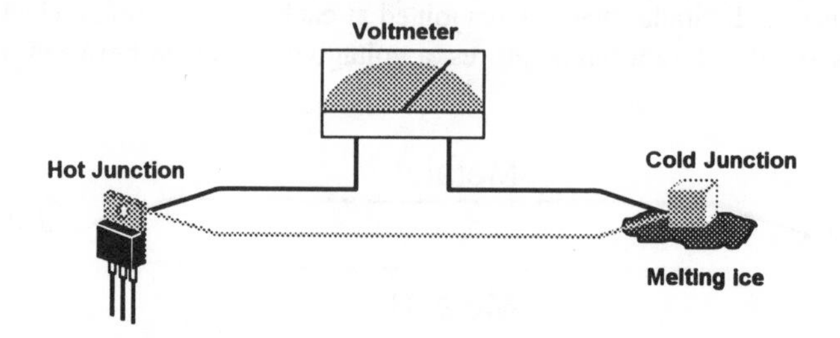

Practical

A semiconductor sensor is used to measure the temperature of the cold junction. It generates a voltage which is equal and opposite to the voltage produced by the cold junction. The voltmeter therefore indicates the temperature of the hot junction in ^{o}C.

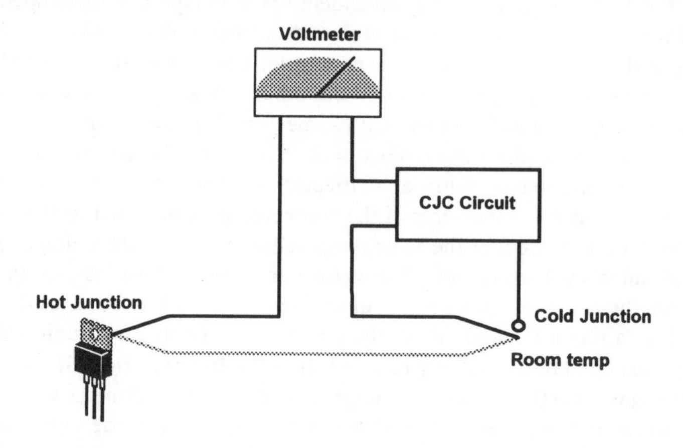

Thermocouple conditioning IC

A semiconductor manufacturer called Analogue Devices has made life very easy by incorporating all the necessary conditioning circuits on one integrated circuit. The circuit is suitable for a type K thermocouple and gives an output of 10mV per ^{o}C. The circuit is suitable for recording temperatures of from -180 to 1100 ^{o}C.

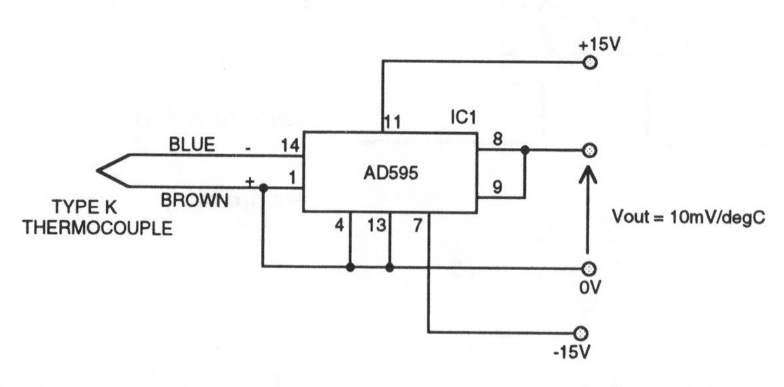

Thermocouple data

ANSI	JUNCTION MATERIALS	ALTERNATIVE NAMES	INSULATION COLOURS	SHEATH COLOUR	TEMP RANGE	TOTAL O/P OVER RANGE	APPROX uV/°C AT 100 °C
B	+ Platinum - 30% Rhodium - Platinum - 6% Rhodium			-	100 to 1600 °C	13.6 mV	1 uV/°C
E	+ Nickel Chromium - Copper - Nickel	Chromel Constantan	Brown Blue	Brown	0 to 800 °C	75 mV	68 uV/°C
J	+ Iron - Copper - Nickel	Constantan	Yellow Blue	Black	-180 to 700 °C	50 mV	46 uV/°C
K	+ Nickel Chromium - Nickel - Aluminium	Chromel Alumel	Brown Blue	Red	-180 to 1100 °C	56 mV	42 uV/°C
R	+ Platinum - 13% Rhodium - Platinum		White Blue	Green	0 to 1600 °C	18.7 mV	8 uV/°C
S	+ Platinum - 10% Rhodium - Platinum		White Blue	Green	0 to 1550 °C	16 mV	8 uV/°C
T	+ Copper - Copper - Nickel	Constantan	White Blue	Blue	-185 to 300 °C	26 mV	46 uV/°C

Source - BS4937,1843

Semiconductor temperature sensor

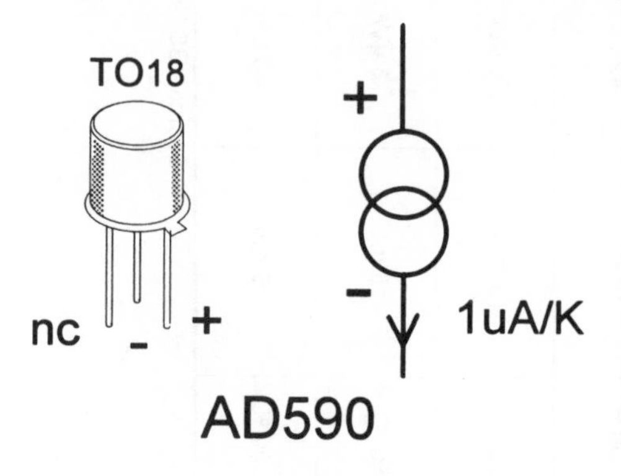

The AD590 is a current source device which gives 1uA per kelvin. It is suitable for the range -50 to +150 °C. When the device is at 0 °C it outputs a nominal 273 uA. This must be allowed for if a reading in °C is required. A suitable circuit to convert the AD590 output to 10mV per °C is shown below. RV1 is adjusted to give 2.73V at the output of IC1. RV2 is adjusted so that the output of IC2 gives 10mV per kelvin. The voltage difference between the two op-amp outputs is therefore 10mV per °C.

AD590 conditioning circuit

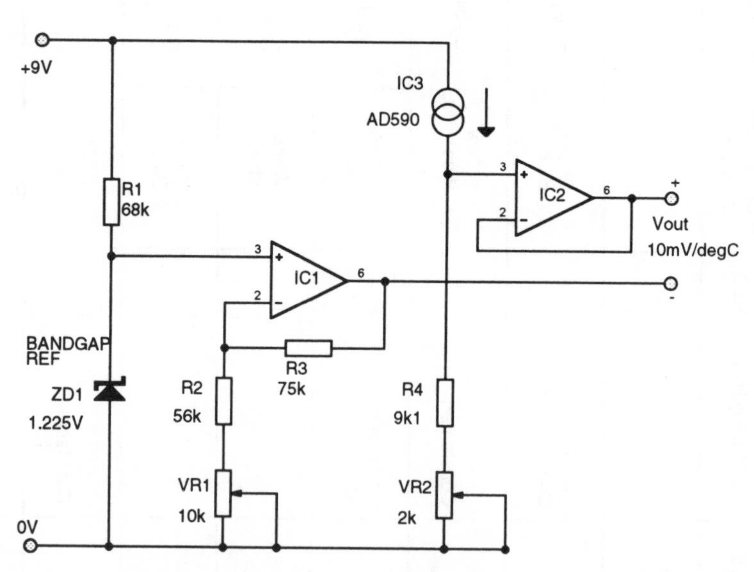

The circuit serves as a useful device to convert a standard digital multimeter to a digital thermometer. The AD590 may be fitted as a probe tip.

Thermistors

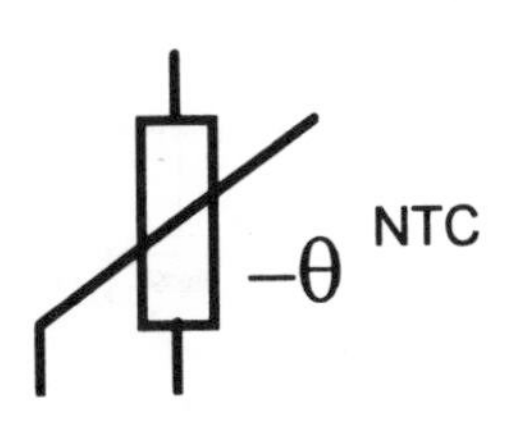

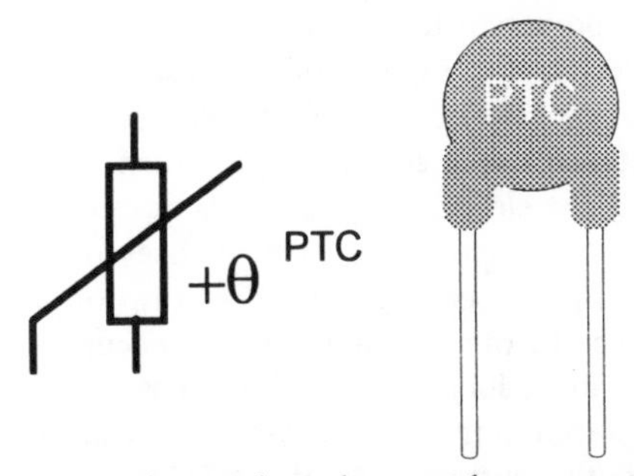

A thermistor is effectively a resistor which is manufactured to have a large temperature coefficient. There are two types available; NTC (negative temperature coefficient i.e. the resistance reduces as temperature increases) and PTC (positive temperature coefficient i.e. the resistance increases as the temperature increases). In practice most thermistors are manufactured from a semiconductor material. PTC thermistors are generally used for control applications such as overcurrent protection and in series with television tube degaussing coils. NTC thermistors are usually used for temperature sensing, although they are occasionally used to reduce current inrush in power supply circuits.

NTC thermistors used for temperature measurement

NTC thermistors have a negative temperature coefficient of between 3% and 6% per °C which is approximately 10 times greater than that of metals (and the opposite polarity). This makes them sensitive devices for sensing temperature and are much lower cost than say platinum resistance thermometers. NTCs have a high resistance compared to other temperature measuring element and so the resistance of connecting leads does not have to be taken into consideration. One disadvantage of NTC thermistors when used for temperature measurement is their non-linearity. Provided the temperature measurement range is limited to between 50 and 100 °C then an NTC may be linearised by parallel and series connected resistors as shown in the circuit below (by permission of Crossland Components). It is important that there is minimal self-heating of the NTC due to the applied voltage across it (see rules of thumb below).

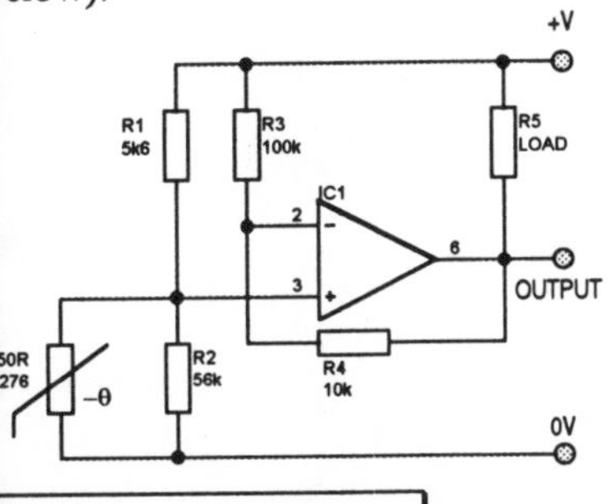

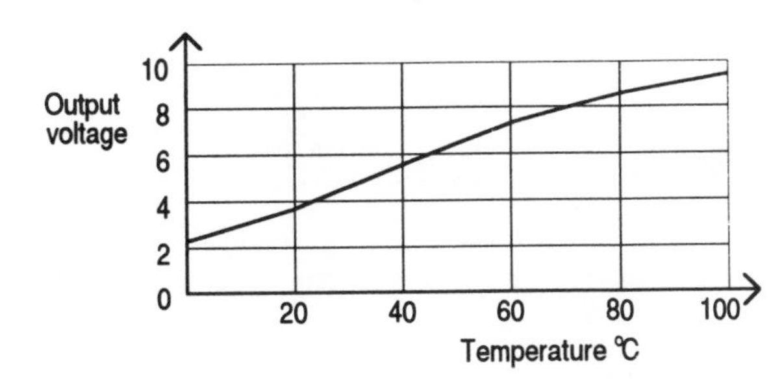

Rules of thumb

- NTC thermistors are generally applicable to measure temperatures between -50 and +300 °C
- At 25 degrees C a 10k ohm NTC will have a temperature coefficient of 4.5% per °C.
- Keep the watts across an NTC below 1mW per °C so that there is negligible self-heating when measuring temperature.

NTC thermistors used for current inrush limiting

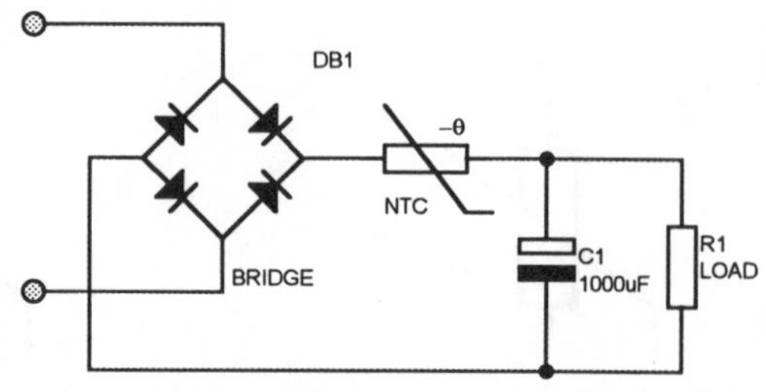

In mains powered dc power supplies, high value electrolytic capacitors are used as reservoir capacitors. When the supply is first switched on, there is a high inrush of current as the electrolytic capacitors charge. This may cause problems such as fuses or current trips operating unnecessarily or interference to other equipment may occur. In order to reduce the initial current, an NTC is placed in the charge path of the capacitor. The cold resistance of the NTC will be hig and limit the current. As the NTC self-heats, its resistance reduces and becomes almos insignificant compared to the resistance of the load.

PTC thermistors used as overcurrent protection

A PTC may be used in place of a fuse to offer protection to equipment should it develop a faul or an overload. During normal operation of the equipment the current drawn from the suppl is not high enough to cause the PTC to self-heat by the amount needed to make it operate. Th resistance stays at a low value. If a fault develops in the equipment and its supply curren increases, then the PTC heats up enough to cause it to increase in resistance. The resistanc increases quite quickly and eventually becomes higher than the resistance of the equipment it i protecting. Most of the supply voltage is then across the PTC. The PTC then keeps itself ho and the current is then reduced to a much lower value. If the supply is switched off and th fault on the equipment clears or is corrected, the PTC will have cooled and reverted to a lo resistance. The equipment will then operate normally when switched on.

6 Circuit concepts

Voltage source

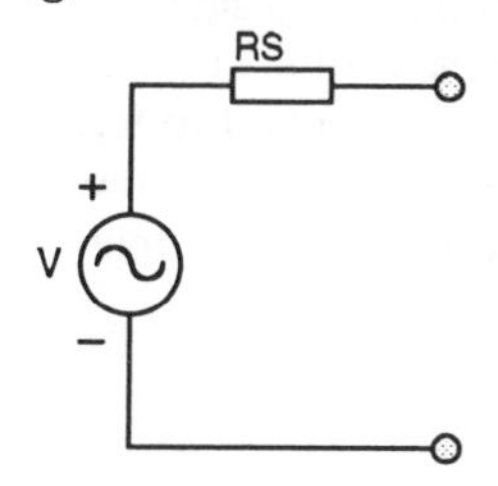

An ideal voltage source (ac or dc) has zero internal resistance Rs so that the output voltage remains constant irrespective of the current drawn from the output.

Example: 3-terminal voltage regulator such as the 7805 is an excellent dc voltage source whose output remains virtually constant for output currents up to about 1A whereupon the device current limits and the output voltage falls.

Current source

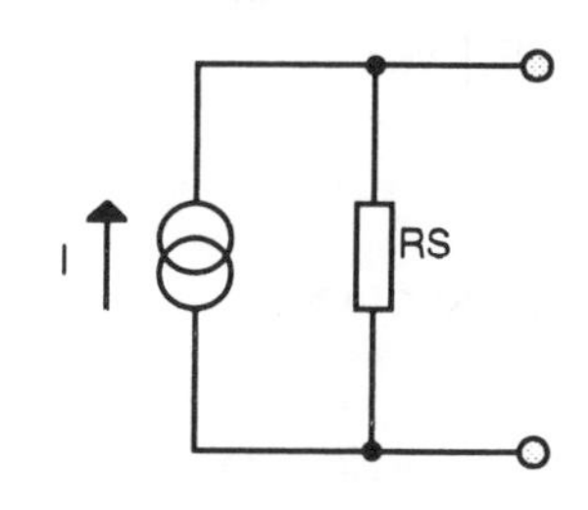

An ideal current source has infinite source resistance so that the output current is constant irrespective of the load resistance applied to the output. A 1A current source may be thought of as a voltage source of say 10,000,000V with a source resistance of 10 megohms. Load resistances of 0 to say 100 ohms will not cause the output current to change significantly.

Example: The drain circuit of a field-effect transistor is a good current source as the drain current is hardly affected by changes in resistance in the drain circuit.

Potential divider

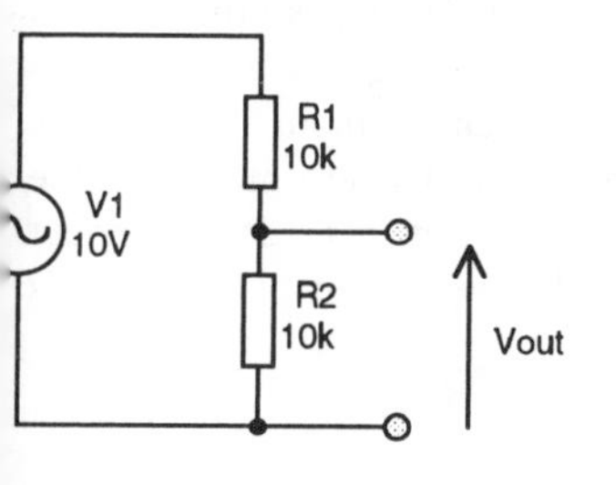

This is formed by two series connected impedances (not necessarily resistors). In the circuit opposite, the voltage Vout is equal to V1 x R2/(R1 + R2) = 10V x 10k / 20k = 5V.

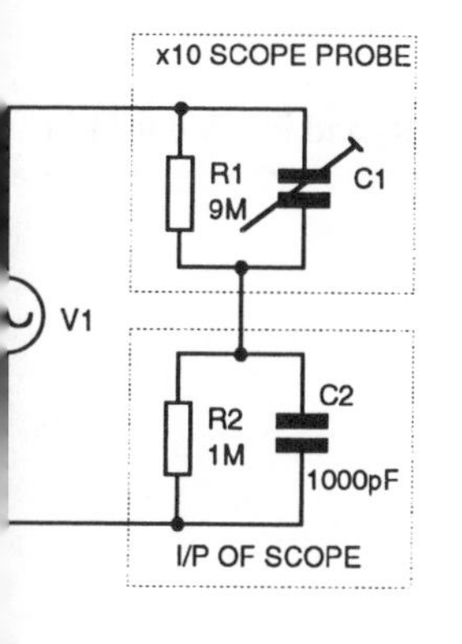

A standard x10 oscilloscope probe uses the potential divider concept. An oscilloscope input impedance is usually standardized at 1M in parallel with 1000pF (R2 and C2 in the figure opposite). A 9M resistor in the scope probe would make a potential divider with the 1M input impedance of the scope and would give the required 1/10 potential division. At high frequencies however, there would be a greater attenuation of the signal V1 due to the input capacitance of the scope C2. To overcome this, a capacitor C1 is fitted in the scope probe which acts as a capacitance potential divider with C2. The probe capacitor C1 is made to be adjustable so it can be made to give an exact attenuation of 1/10 with C2 at high frequencies.

Voltage gain and current gain

The voltage gain aspect of an amplifier is exhibited from input to output; i.e. a signal voltage is applied to the input and a greater amplitude version of the signal appears at the output.

The current gain aspect of a power amplifier is usually exhibited from output to input. The current gain will not be seen until a load is connected which draws current from the output. The input of the amplifier will then demand a current which is equal to the output current divided by the current gain.

Diode pump

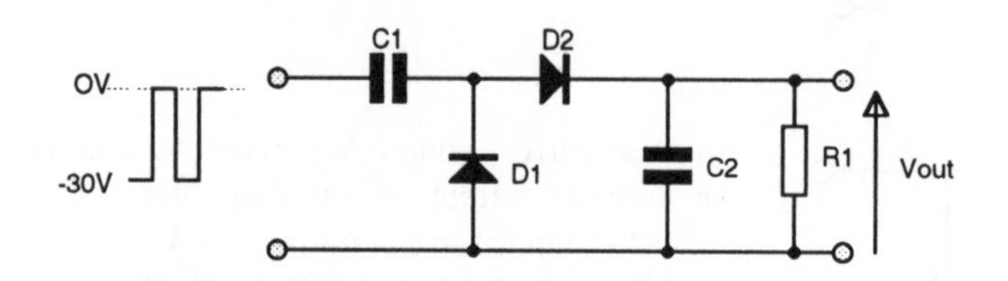

THE OUTPUT VOLTAGE IS SET BY C_1 AND THE FREQUENCY.
THE OUTPUT RIPPLE IS SET BY C_2.

The diode pump gives a dc output which is dependent upon the peak to peak ac input voltage and its frequency. Any dc component on the input waveform is ignored. It may be used as a crude frequency to voltage converter or to create a low voltage dc power supply from a higher voltage ac supply such as the 240V mains supply.

On each cycle of the input signal, charge is collected on C1 and then dumped into C2. The charge on C2 is constantly leaking away through R1. Equilibrium is reached when the charge gained by C2 from C1 equals the charge lost by C2 into R1. If C2 is made large so that there is a small ripple present at Vout, then Vout may be considered constant. The charge gained by C2 is the charge lost by C1 on each cycle of the input waveform. When the input waveform is at -30V, then C1 is charged to 30V. When the positive edge of the input waveform occurs, the right hand plate of C1 would be raised by 30V if D2 was not in circuit. D2 does conduct however and charge flows from C1 to C2. D2 turns off when the right hand plate of C1 is discharged to a voltage equal to Vout.

Charge lost by C_1	**$= C_1 \times$ change in V across C_1**
	$= C_1 (30 - V_{out})$

Charge lost by C_2 during one period T of the input square wave is
$= I\,T$ (where I is the current drawn by R_1 and is = Vout / R_1).

Charge lost = charge gained

$$V_{out} T / R1 = C1 (30 - V_{out})$$

$$V_{out} = \frac{30}{1 + 1 / (fR1C1)}$$

Where f = 1/T.

Wheatstone bridge

The Wheatstone bridge consists of four resistors connected as two potential dividers in parallel. It may be used to accurately determine the value of an unknown resistor. R1 and R2 are precision resistors of the same value. The unknown resistor is placed at R3 and R4 may be a decade box of precision resistors. When R4 equals R3 then the voltage at point X equals that at point Y and there is no output voltage of the bridge and it is said to be balanced. If a very sensitive ammeter is connected at the output of the bridge then in order to determine the resistance of the unknown, the decade box R4 is adjusted until zero current flows in the ammeter. The unknown resistor is then equal to that value on the decade box.

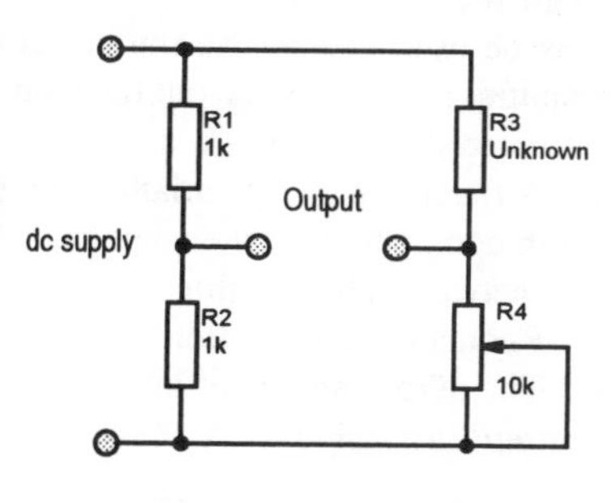

In general for a Wheatstone bridge:

$$\frac{R1}{R2} = \frac{R3}{R4}$$

A Wheatstone bridge may also be used to accentuate small changes in the resistance of a sensor such as a strain gauge. If the strain gauge is connected in position R3 and R4 adjusted until the bridge is balanced, then small changes in the strain gauge will be seen as a significant deflection of the ammeter.

Balanced circuit

This term occurs most frequently in telephony or audio systems and is used to reduce the effects of noise when transmitting signals on long cables. Whenever signals are transmitted over long cables, unwanted noise is picked up from interference sources such as power lines, electrical appliances and machines. A balanced system uses two conductors which are arranged to be electrically identical with respect to earth. Consequently electrical noise will be generated equally in each conductor. The signal receiver is designed to only amplify the voltage difference between the two conductors. The signal receiver is also carefully designed so as to present the same electrical impedance to each of the conductors. The cable used in balanced systems usually comprises a twisted pair of conductors in a outer screened sheath. The twisting helps to expose each conductor equally to any electrical noise which may be present. The receiver often takes the form of a differential amplifier.

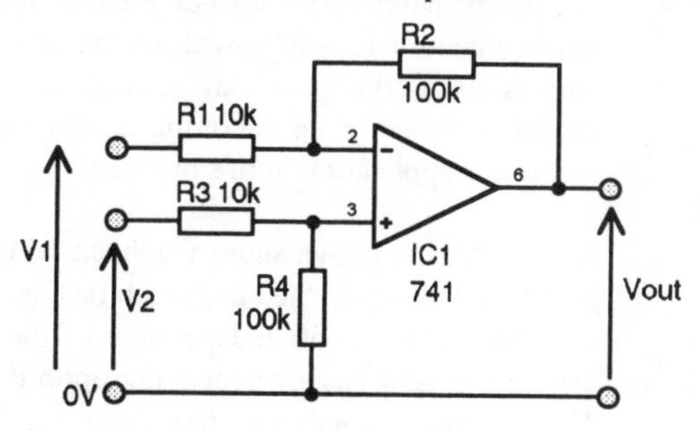

Negative feedback
May be used to improve the performance of amplifiers. A fraction of the output signal of a amplifier is subtracted from the input signal before it is applied to the input of the amplifier. Th resulting improvements are:

- A reduction in the variations in gain of the amplifier due to temperature, age and variation in components in the amplifier.
- Increase in bandwidth.
- Reduction in distortion.
- The output and input impedances are altered.

Example 1: single NPN transistor amplifier

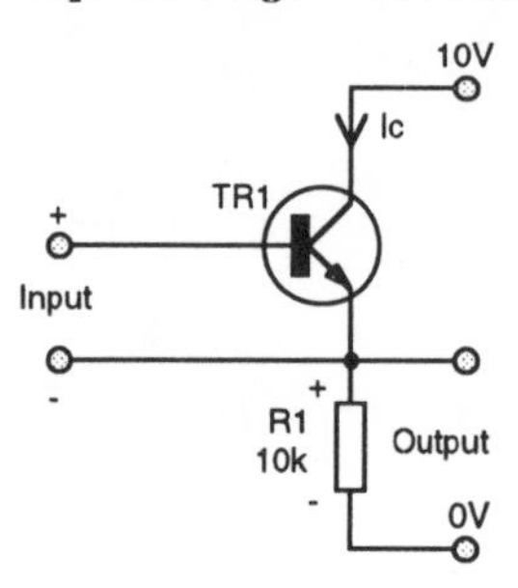

The input voltage is applied between the base and the emitter o the transistor. The input impedance of the transistor is very lov and depends upon the input voltage. Input voltages of no mor than a fraction of a volt may be applied. The output of th transistor is its collector current Ic. This also flows out of th emitter. In order to produce an output voltage of the amplifier a resistor R1 is used to convert the output current to a voltage The resulting voltage amplifier is very non-linear and drift severely with variations in temperature. If a different transisto is used, the amplifier's performance will be changed dramatically

With negative feedback (the emitter follower)

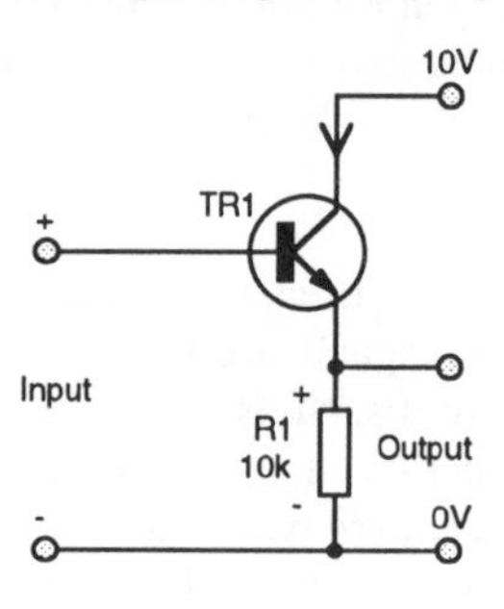

The input voltage to the total stage is now applied between th base of TR1 and 0V. Consequently the input to the transisto (between its base and emitter) is equal to the input voltage to th stage minus the output voltage. The entire output voltage ha been fed back in series with the input of the transistor and it is o the opposite polarity to the input to the stage. In this circui 100% negative feedback has been employed. The result is a stage which has a voltage gain of 1 but is very linear an insensitive to temperature changes and different transistors More importantly, the input impedance has been increased an the output impedance has been reduced.

Example 2: operational amplifier

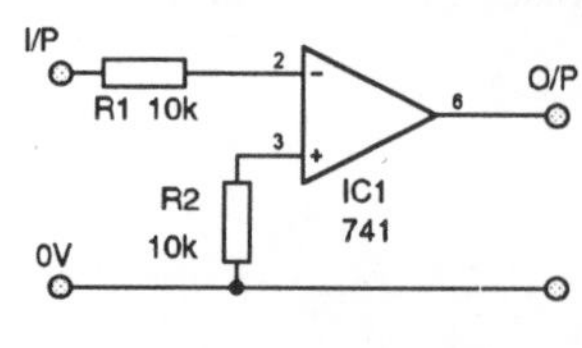

IC1 is connected as an inverting amplifier. It has a voltage gai of approximately 200,000 but a bandwidth of only 10Hz. Th voltage gain drifts wildly with temperature and if a different op- amp is used, the gain may change by up to 10 times. Th amplifier is said to be operating on open-loop and is never use in a linear application in this manner.

With negative feedback

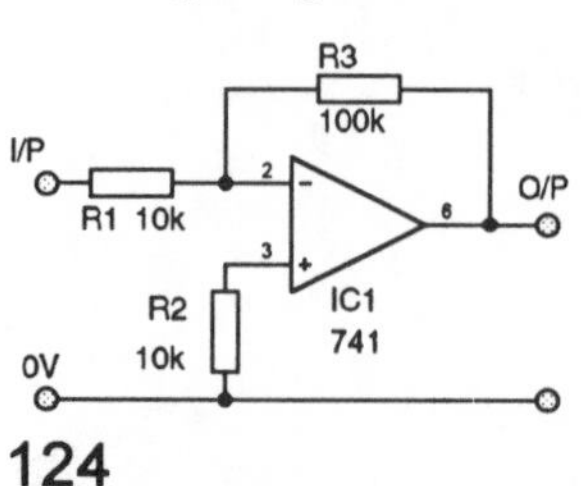

R3 provides negative shunt feedback which reduces the voltag gain to 10 however the bandwidth has been increased to 100kHz and the gain is almost independent of the operational amplifier' gain and is now largely dependent upon the resistors R1 and R3 The output impedance has been reduced, but the inpu impedance is now R1 due to the virtual earth effect.

Positive feedback
Is sometimes referred to as regenerative feedback. A fraction of the output signal of an amplifier is added to the input signal and in phase with the input signal before it is applied to the input of the amplifier. Positive feedback is less common today compared to when amplifying devices were more expensive. Positive feedback was used to increase the gain of an amplifying stage or stages, however it has the drawback that the total circuit could become unstable and self-oscillate.

Positive feedback is more commonly used in comparator circuits where it is employed to eliminate the effects of noise on an analogue signal which is being compared with a threshold level.

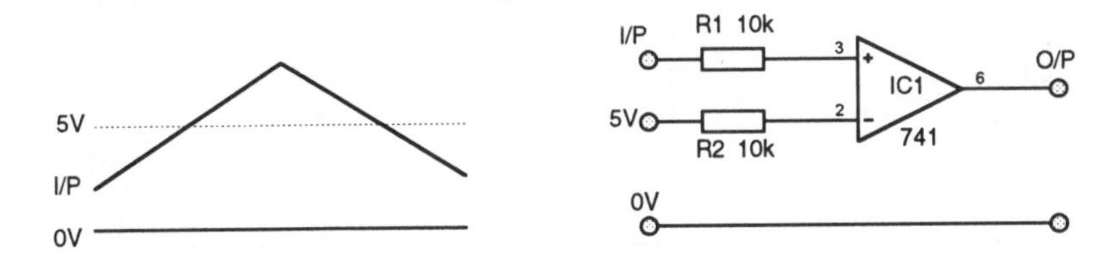

In the circuit above, the input voltage is a slowly varying dc signal which may represent the temperature of a room. IC1 is connected as a comparator which compares the dc input signal with a 5V reference. When the input voltage exceeds 5V, then the output of IC1 switches from negative to positive saturation. If the input signal contains noise, then the input signal may vary above and below the 5V reference when it is in the region of 5V. This will cause the output of IC1 to rapidly switch between positive and negative saturation which may not be desirable if the op-amp is controlling a high power device via a relay. The relay would be made to switch in and out very quickly which could lead to premature contact wear.

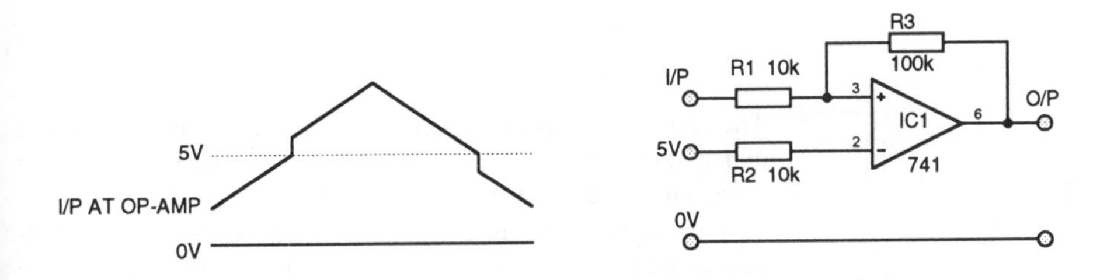

In the modified circuit above, positive feedback has been added via R3. The input to the op-amp is now the addition of the input signal and approximately 10% of the output signal. When the input signal reaches 5V, the output of the op-amp switches to positive saturation and adds to the input signal. The input to pin 3 is consequently elevated by the positive feedback and quickly raised above the 5V threshold level. The effects of any noise on the input signal are therefore reduced. A similar phenomenon occurs when the input signal is falling. The net result is that the circuit switches on a higher input signal when it is increasing to that when it is decreasing. The effect is often referred to as hysteresis.

Miller effect

In figure (a) below, the lower plate of C1 may be considered to be stationary and the upper plate is being varied in voltage by V1. As V1 varies, a current flows to charge and discharge the capacitor according to the equation **i = C1 dV/dt.** The voltage generator "sees" a capacitance of C1. When a second voltage generator V2, equal in magnitude, frequency and phase to V1, is inserted as shown in figure (b), both plates of the capacitor are varying in sympathy. No current will flow from V1 and voltage generator V1 "sees" zero capacitance. If V2 is reversed as in figure (c), then the magnitude of the voltage across C1 is twice what it was in figure (a). The voltage generator V1 actually "sees" 2 x C1.

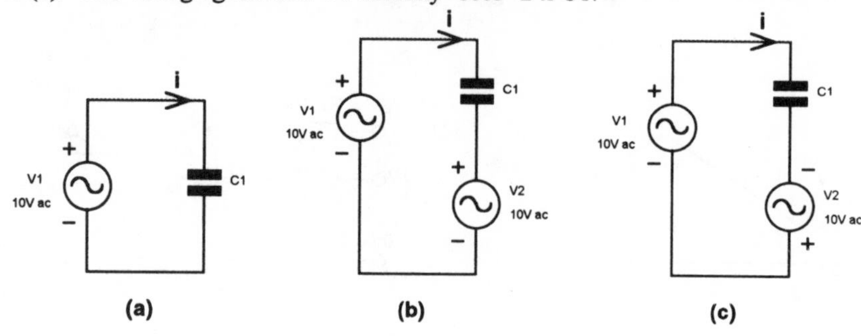

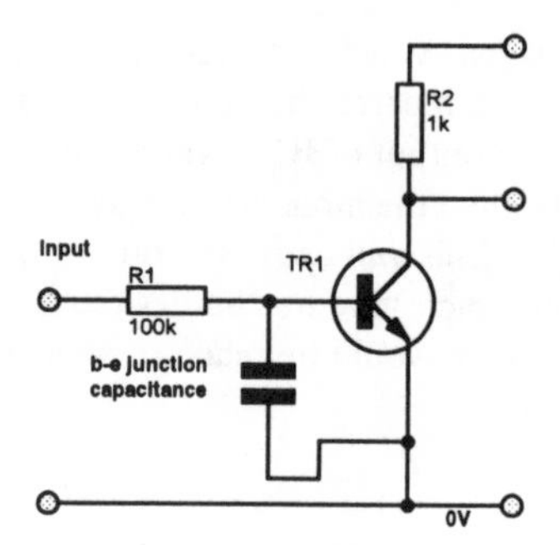

All semiconductor junctions have capacitance. This cannot be entirely avoided during manufacture. The base -emitter junction of a bipolar transistor for example will have capacitance and when the transistor is used in the common emitter mode, the base -emitter capacitance will cause current to be bypassed to 0V at higher frequencies when its reactance is in the same order as the base resistor. This will reduce the gain of the stage.

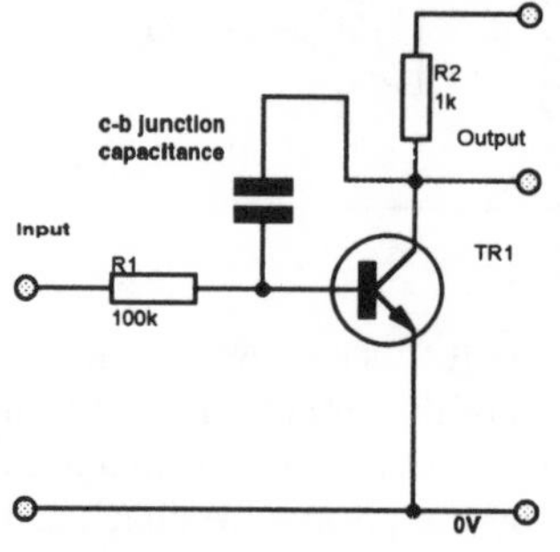

The transistor's collector -base junction also has capacitance and if the voltage at the collector was not varying then the capacitance "seen" at the input would simply be the collector - base capacitance because the plate of the capacitor which is connected to the collector may be considered to be at 0V (ac wise). The voltage at the collector is varying however as this is the output voltage of the circuit. The capacitance "seen" at the input is actually magnified by an amount equal to (1 + the voltage gain of the circuit). This is the Miller effect.

Bootstrapping

This could be considered to be the opposite of the Miller effect. In the previous section it was seen that the Miller effect increased the effective value of the collector -base capacitance because the collector was moving in antiphase to the input signal. The effective value would have been reduced if the collector had been moving in phase with the input signal. The concept would be the same if the parameter in question were resistance rather than capacitance.

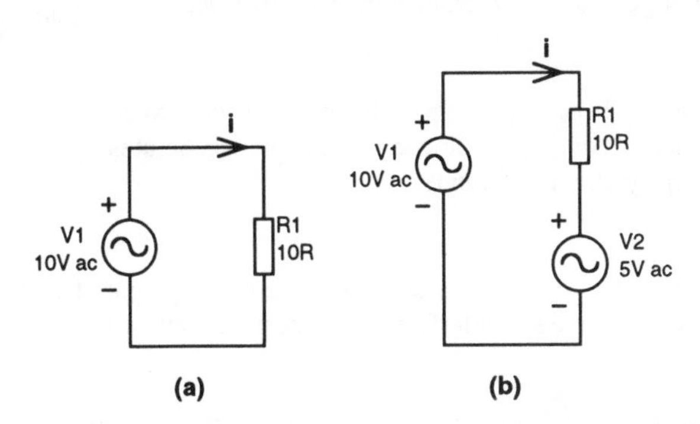

In diagram (a) above, the ac voltage source V1 "sees" R1 as being 10 ohms. In diagram (b) the ac voltage source V2 is exactly the same frequency and phase as V1. The current i is therefore 10V - 5V divided by 10R = 0.5A. V1 "sees" 20 ohms. The voltage source V1 is unaware that the other end of R1 is connected to V2 and so V1 "sees" 20 ohms (10V / 0.5A = 20R) The other end of R1 is being bootstrapped by V2. Bootstrapping increases the effective value of a resistor (or in fact any impedance). The technique may also be used to reduce stray capacitances in a circuit.

The concept of bootstrapping was given its name by the bizarre idea that a person could defy gravity by pulling himself up by his own bootstraps (bootlaces in the UK) thereby reducing his own weight by however hard he pulled.

Balanced output

In audio power amplifiers employing semiconductor output devices the maximum output power available into a loudspeaker is limited by the dc power supply voltage. This is because the output ac voltage of the amplifier cannot swing higher or lower than the power supply rails. In practice, the output voltage is limited to a few volts less than the supply rails because of the saturation voltages of the output transistors. The output power can be increased approximately four times by employing a second output stage which is in antiphase and connecting one terminal of the speaker to one output and the other terminal to the second output stage.

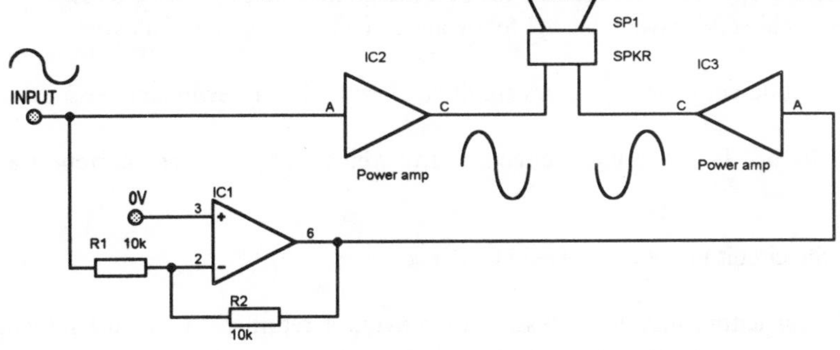

7 Electromagnetic compatibility

Definition

Electromagnetic compatibility (EMC) is the subject which considers how an item of electrical or electronic equipment may:

1. interfere with other electrical or electronic equipment due to it generating electromagnetic interference,

   This is called **the emission aspect.**

2. be interfered with by other electrical or electronic equipment or phenomenon which generates electromagnetic interference.

   This is called **the immunity aspect.**

Depending upon the type of equipment, standards are being prepared or have been published which detail the maximum levels of interference for both aspects of EMC.

Electromagnetic interference generated by equipment may be transmitted to other equipment:

1. Via interconnecting cable e.g. the mains supply.

   This is called **conducted interference**

2. Through space by electromagnetic waves.

   This is called **radiated interference**

Up to a frequency of approximately 30MHz, interference signals are usually propagated by conduction through interconnecting cables. (This is sometimes called galvanic coupling). Above 30MHz, interference signals are usually propagated by radiation.

Design guidelines to reduce emission

In general it is rate of change of current which generates the severest interference. High switching speed semiconductors are the greatest source of conducted and radiated interference. If high switching speed is not required, the best design maxim is to reduce di/dt.

If fast switching is unavoidable the following guidelines should be followed:

- Ensure that de-coupling capacitors are fitted close to the fast switching devices.
- Avoid "loops" in the printed circuit board which carry the fast current transients or "edges".
- Mount the circuit inside a screened metal box.
- Class Y capacitors may be necessary to provide a return path for interference currents flowing from the circuit to the metal box via stray capacitances.

- Fit a class X capacitor across the incoming mains supply to de-couple high frequency components of the input current demand.
- Add series inductors to further encourage high frequency components of current to flow from de-coupling capacitors rather than from the power supply.

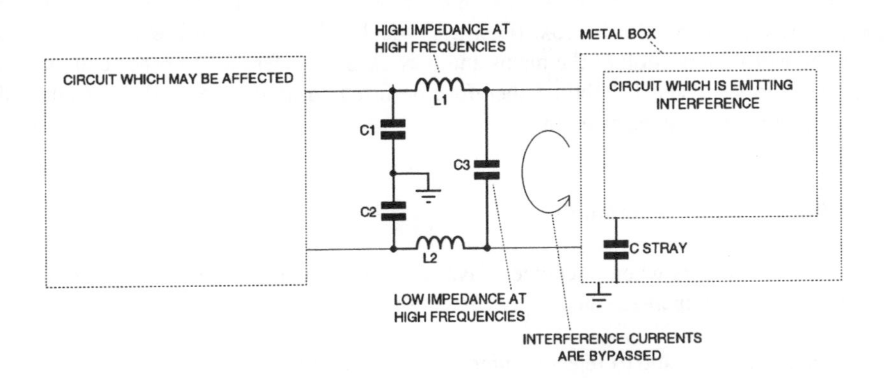

Example showing methods of reduction of emission

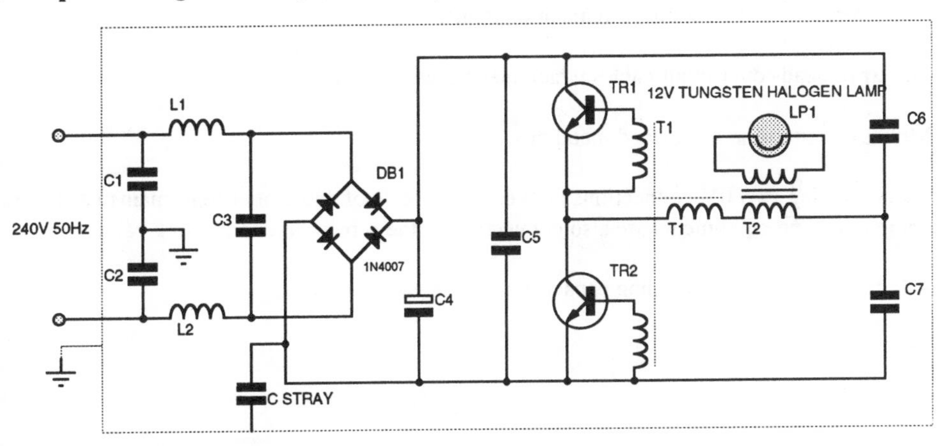

The above circuit is a half-bridge inverter which runs at approximately 30kHz and is used to transform mains power down to 12V to operate a tungsten-halogen lamp. The two transistors TR1 and TR2 generate interference due to the fact that they are switching current at high speed. In order to reduce emission, the following guidelines apply:

- C5, C6 and C7 are high frequency de-coupling; they should be a low inductance type and should be mounted close to TR1 and TR2.
- Avoid creating loops in the printed circuit board in the tracks that carry the current from C5, C6 and C7 to TR1 and TR2.

- Fit a class X de-coupling capacitor C3 to further prevent the high frequency harmonics of current (due to the fast "edges" of current) from being drawn from the mains.

- Insert chokes L1 and L2 to introduce a series impedance to fast "edges" of current. L1 and L2 further encourage the high frequency harmonics of current to flow through C3.

- If the circuit is mounted in a metal box, the box must be earthed for safety reasons. The box will then reduce the amount of radiated interference by screening. However the circuit will have stray capacitance to the box, (C stray) which will cause interference currents to flow into the earth connection of the mains and may cause interference to other equipment. Class Y capacitors C1 and C2 should therefore be fitted so as to bypass the current and prevent it flowing into the mains earth.

Design guidelines to improve immunity

- Use lower value resistors where a choice is available (for example, when biasing the inputs of op-amps). This will increase the supply current of the circuit, however.

- Use a differential input stage to reject common-mode interference.

- Fit de-coupling capacitors on susceptible inputs.

- Use screened cable to connect items of equipment.

- Fit ferrite beads over input cables which may be susceptible.

- Mount the circuit in an earthed metal box.

- Avoid making two 0V connections between two items of equipment (e.g. mains earths are commoned and equipments are also connected via the screen of a connecting cable.

- Add further dc rail de-coupling capacitors at high gain stages.

8 Sound

Sound waves are longitudinal wave vibrations usually in air, i.e. air vibrates in the same direction to which the sound wave is travelling. The audible frequency range is approximately 20Hz to 20kHz depending upon the individual. The upper limit is related to age; a young person's hearing may extend to 20kHz, however, an old person may only be able to hear up to 6kHz or less.

The dynamic range of sound is measured in decibels where 1dB = 10 times the logarithm of the ratio of the power levels of the loudest sound to the quietest sound. The dynamic range of sound levels in a classical music performance may be from 30 dB during very quiet passages to 100 dB during the 1812 Overture, (a range of 70dB).
The range of sound levels occurring in everyday life is shown in the table below.

DECIBELS	PRESSURE		EXAMPLES
120	20Pa	THRESHOLD OF PAIN	JET AIRCRAFT AT 500 FT
110			POPULAR MUSIC BAND
100	2Pa		INSIDE A TUBE TRAIN
90			BUSY STREET
80	0.2Pa		WORKSHOP
70			NOISY OFFICE
60	0.02Pa		NORMAL CONVERSATION
50			QUIET OFFICE
40	2mPa		RURAL HOUSE
30			QUIET CONVERSATION
20	0.2mPa		WHISPER
10			STILL NIGHT IN THE COUNTRY
0	20uPa	THRESH. OF HEARING	SOUND-PROOF ROOM

The decibel is actually a ratio:

1dB = 10 x log P1/P2 (ratio of two power levels)
1dB = 20 x log V1/V2 (ratio of two voltage levels)

The above table is based on 0dB being a sound pressure of 20uPascals = 200udynes/cm^2
Other decibel systems in use are:

dBm: (based on 0dB = 1mW)
dBu: (based on 0dB = 0.775V rms)
dBV: (based on 0dB = 1V rms)
dBW: (based on 0dB = 1W)

Glossary of audio terms

Bandwidth
The range of frequencies over which an audio system is designed to operate. In general the bandwidth is the range between the two frequencies where the gain is 3dB lower than the gain in the mid-range frequencies.

Compression
The dynamic range of a signal is compressed i.e. quiet sounds are made louder, loud sounds are made quieter. The gain of a compressor is inversely dependent upon the input signal. It is used to enable signals which have large variations in sound level (e.g. a symphony orchestra) to be processed by a system which has a limited range of levels (e.g. an a.m. radio transmission system).

Current gain
The current gain aspect of a power amplifier is usually exhibited from output to input. The current gain will not be "seen" until a load is connected which draws current from the output. The input of the amplifier will then demand a current which is equal to the output current divided by the current gain. So the current gain is the output current divided by the input current.

Decibel
See previous page.

Graphic equalizer
This is a comprehensive tone control system where the audio frequency range is divided into specific bands and each one may be controlled by a potentiometer which enables the frequencies in the band to be boosted or attenuated. Each band is centred on one particular frequency which receives the greatest boost or cut with reducing effects to the frequencies either side of the centre frequency at a rate which depends upon the design of the equalizer.

Harmonic distortion
This is where additional frequency signals are introduced by an audio system which are multiples of the frequencies of the true signals being processed. The distortion is introduced because of non-linearities in the audio system.

Limiter
A limiter is an electronic circuit which processes a signal so that levels below a pre-set threshold are unaltered, but signals exceeding the threshold are reduced in dynamic range to prevent them exceeding a certain maximum value. It is used to ensure that clipping does not occur in a subsequent amplification system.

Pink noise
This has equal noise power per octave of the audible frequency range. It is used to evaluate graphic equalizers as equal amplitudes of noise are applied to each band.

Power amplifier
This supplies the necessary voltage gain and current gain to enable a specified power to be supplied to the specified load (e.g. a loudspeaker). A typical audio power amplifier will give its rated power output into say a 4 ohm load with an input signal of 0.775V (0dBu).

Pre-amplifier

A pre-amplifier is used:

- to match the input transducer to the later stages of the main amplifier,
- to correct any frequency characteristic that has been made to the signal,
- to modify the frequency characteristics to suit listening conditions,
- to provide sufficient voltage gain and current gain to drive the input of a power amplifier.

Signal to noise ratio

This figure is the ratio in decibels of the average signal level in volts (Vs) to the total noise voltage (Vn)

S/N ratio = 20 log Vs/Vn.

Voltage gain

This is usually exhibited from the input of an audio system to the output. If a signal voltage is applied to the input, then an audio system will have voltage gain if the output voltage is larger than the input voltage. In general the voltage gain of a system does not take into account the ability of the output to drive a load.

White noise

This is equal noise power per Hz over the complete audio frequency range.

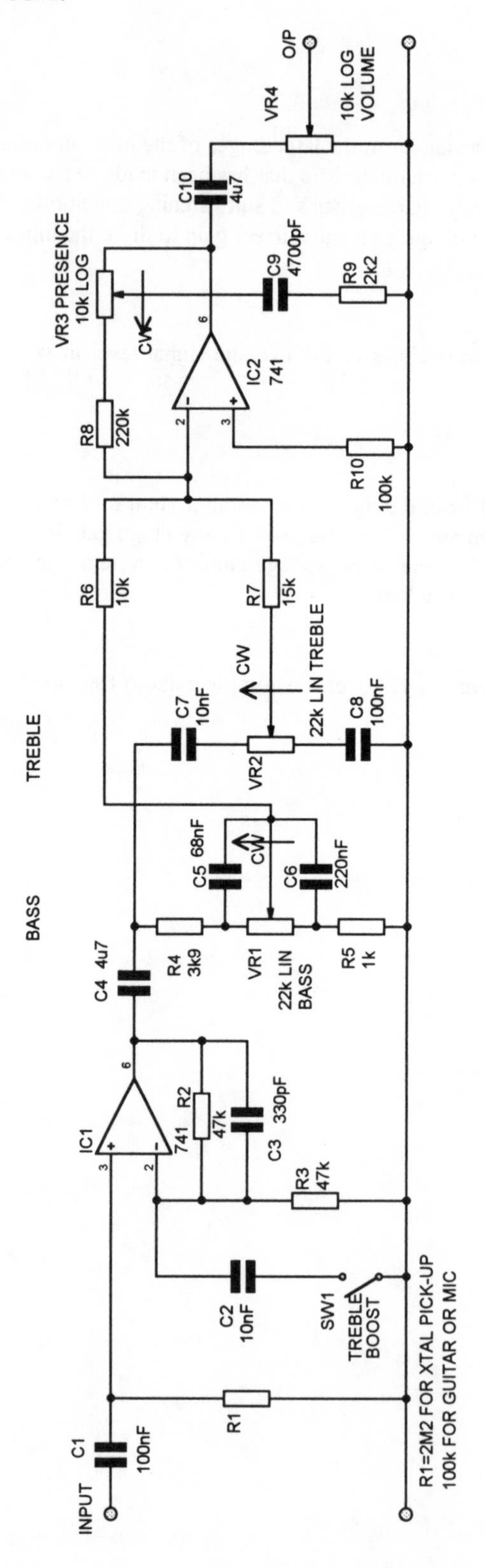

A TYPICAL AUDIO FREQUENCY PRE-AMP

Power amplifiers

There are many different configurations of power amplifiers, however, there are six main stages or sections which are common to most power amplifier designs.

- Power modulators - output circuit (TR3 to TR6).
- Biasing circuit determining the class of operation (X).
- Biasing circuit determining the quiescent conditions (R1, R2).
- AC gain determining circuit (R3, R4, C2).
- Bootstrap circuit or similar to improve power efficiency (C3).
- Coupling of the load (C4).

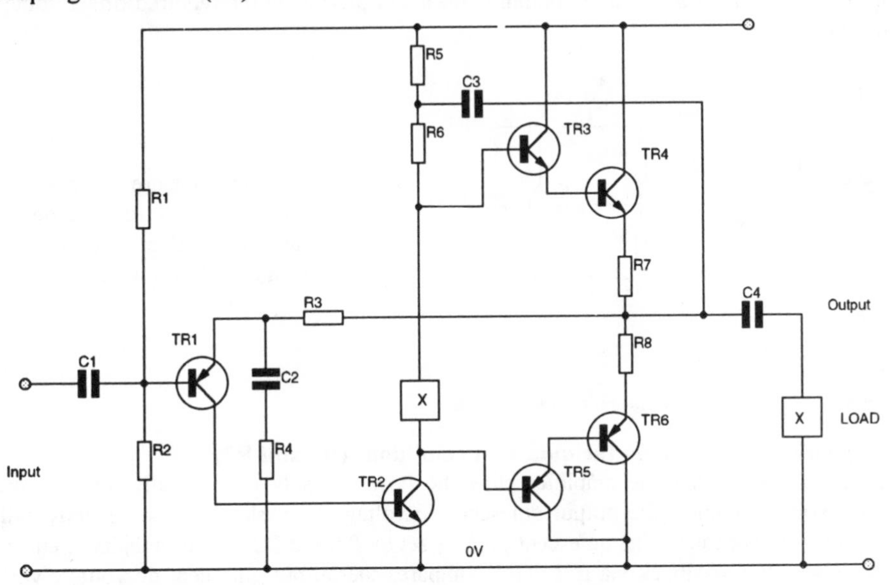

Power modulators (TR3 to TR6)

These are the output devices and give most of the current gain and very little voltage gain. The configuration shown above is a complementary emitter follower. Other possible configurations are shown below.

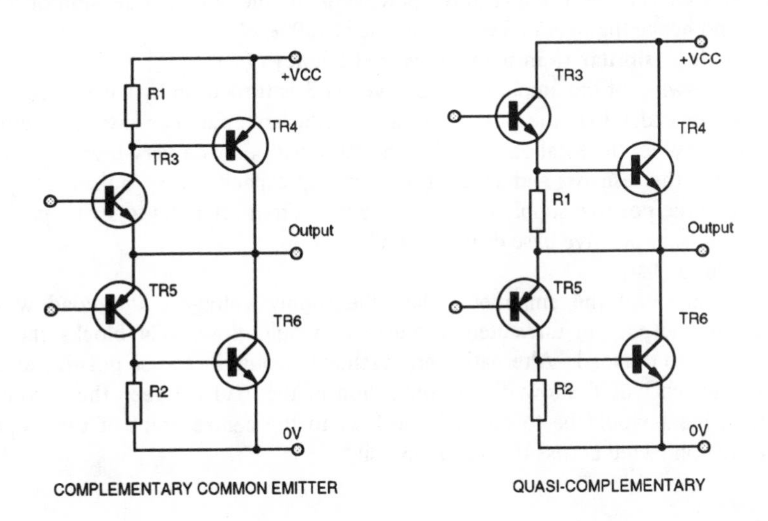

Sometimes Darlington pairs are used to increase the current gain, but more recently power MOSFET devices are used for the output devices as they provide a very high current gain and are voltage driven.

Biasing circuit determining the class of operation (X)
In all cases, the class of operation depends upon the bias voltage between the two input bases. The transfer admittance (collector current out for base- emitter voltage in) characteristic of a transistor is very non-linear at very low voltage levels and an offset voltage is required between the two input bases of the power devices to overcome the non-linearity which is termed cross-over distortion. Two of the more common means of producing the required offset voltage are shown below.

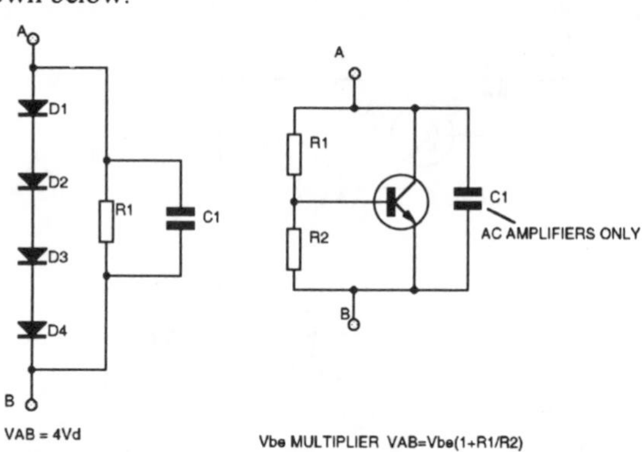

These circuits are often in thermal contact with the output devices so that the offset voltage tracks changes in Vbe's with temperature.

Biasing circuit determining the quiescent conditions (R1 and R2)
In order to ensure that the amplifier clips the power supply rails equally for positive and negative voltage swings, the output quiescent dc voltage level should be set exactly half way between the power rails. The quiescent point is set by R1 and R2 and its stability is ensured by 100% dc negative feedback via R3. TR1 compares the output quiescent dc voltage with that set by R1 and R2 and adjusts the output accordingly via the intermediate stages. The function of TR1 is often accomplished by a differential amplifier such as a long-tailed pair or an op-amp.
AC gain determining circuit (R3, R4, C2)
Power amplifiers usually require voltage gains in the order of 10 to 50. The overall negative feedback for ac signals is set to the required percentage by the potential division of R3 and R4. C2 blocks dc and hence the overall feedback for dc is 100%.
Bootstrap circuit or similar to improve power efficiency (C3)
When the output swing of the amplifier is positive, TR3 is turned on by base current via R5 in series with R6. In order that the output voltage of the amplifier can swing as close to the positive rail as possible, the total value of R5 and R6 must be made very low. This may cause over-dissipation in the resistors and TR2. The bootstrap capacitor C3, "hoists" the junction of R5 and R6 above the positive supply rail each time the output attempts to swing positive. This provides the necessary positive base drive for TR3.
Coupling the load (C4)
The quiescent output of the amplifier is half the supply voltage. If a load was directly connected to the output, an unwanted dc current would flow. C4 blocks the dc whilst allowing ac power to the load. Alternative arrangements would be to use positive and negative dc power rails which would allow direct connection of the load between the output and 0V. Another arrangement would be to connect the load to the centre point of two capacitors in series which are connected across the dc supply rail.

Electronics and music

Musical notes are made up from combinations of sine-wave vibrations. Some musical instruments such as the flute give almost pure sine tones. In addition to the note being played, most instruments generate harmonics or overtones which are usually related to the fundamental note by exact multiples of its frequency. The human ear is able to analyse the different harmonic pattern of each instrument to tell them apart even though they may be playing the same note. Fourier analysis is ideally applicable to the sound of musical instruments.

Musical instruments also differ in the characteristics of attack and decay of the sound. Plucked instruments have a very fast attack; bowed instruments take longer to "sound"

- The audible frequency range is approximately 20Hz to 20kHz depending upon the individual.
- A doubling in frequency of a note is termed an octave in music.
- In western music, the octave is divided into 12 notes.
- On keyboard and fretted instruments, the interval between each note is the 12th root of 2. This is termed the scale of equal temperament as there is the same frequency ratio between one note and the next. This scale does not give the purest harmonies, however. "Analogue" instruments such as a violin can play the true harmonies and sound more pleasing to the trained musical ear.

Electronic effects in popular music

Fuzz

The waveform of the instrument, (usually an electric guitar) is deliberately clipped or square wave tones are generated which are equal in frequency to the note being played. The waveform may be clipped in the pre-amp stages of the amplifier by connecting back to back zener diodes across the feedback resistor of an op-amp stage. Fuzz was first popularized by the Rolling Stones in their recording of "(I Can't Get No) Satisfaction".

Wah-wah

The instrument is played through a narrow bandpass filter whose centre frequency may be shifted up and down the frequency spectrum usually by a foot pedal device.

Phasing or chorus effect

The signal from the instrument is delayed and then mixed with the undelayed signal. Some frequencies are shifted in phase by 180 degrees and are cancelled out. A notch comb filter is created having notches an octave apart. If the delay is varied slowly and in a ramp fashion, the phasing effect is produced which is similar to the "jet-plate" effect heard when listening to a short-wave radio station which is fading and increasing in strength due to the ground wave and sky waves adding and cancelling. This effect was first popularized by the band The Small Faces in their recording "Itchycoo Park".

If the delay is varied in a more random manner, the chorus effect is produced where the impression of many instruments playing simultaneously is created. This may be explained by considering many violins playing the same note simultaneously; the waveforms will vary in phase and some harmonic frequencies will add and some will cancel at random.

Tremelo
Tremelo is amplitude modulation of the instrument's sound and it is created electronically by varying the gain of the amplifier in a low frequency sinusoidal manner at about 2 to 5 Hz. Certain cultures in the world use tremelo in the voice when singing.

Vibrato
Vibrato is frequency modulation and it is used by singers and musicians in western cultures. It is created by varying the pitch of the note up and down around the true pitch of the note being played or sung. It is created electronically in say an electronic organ, by varying the frequency of the oscillator which is generating the note being played.

Reverberation
This occurs naturally in large halls when reflections from walls and ceilings of the sound of the performers arrive later in time than the direct sound. There is a multitude of reflections which all blend together to give a pleasing, sustaining effect to the sound. It may be created electronically by the use of acoustic spring delay lines or plates but more recently is created by delaying the sound using semiconductor devices.

Echo
Is similar to reverberation but the sound reflections are fewer and are separately distinguishable. It may be created electronically by using magnetic tape-loop delay systems but more recently by using semiconductor devices.

			Scale of equal temperament		Pure scale		
MUSICAL NOTE	TONIC SOL-FA	MUSICAL NAME	FREQUENCY Hz	FREQUENCY RATIO TO C	FREQUENCY Hz	FREQUENCY RATIO TO C	MUSICAL INTERVAL
C	Doh	TONIC	261.63	1	261.63	1	UNISON
C#	De	-	277.19	1.059	-	-	-
D	Ray	SUPER-TONIC	293.67	1.122	294.33	9/8 = 1.125	2nd
D#	Re	-	311.13	1.189	313.96	6/5 = 1.2	Minor 3rd
E	Me	MEDIANT	329.63	1.26	327.04	5/4 = 1.25	3rd
F	Fah	SUB-DOMINANT	349.23	1.335	348.84	4/3 = 1.333	4th
F#	Fe	-	370	1.414	-	-	DIMINISHED 5th
G	Soh	DOMINANT	392	1.498	392.45	3/2 = 1.5	5th
G#	Se	-	415.31	1.587	418.61	8/5 = 1.6	AUGMENTED 5th
A	Lah	SUB-MEDIANT	440	1.682	436.05	5/3 = 1.666	6th
A#	Le	-	466.17	1.782	470.93	9/5 = 1.8	7th
B	Te	LEADING NOTE	493.89	1.888	490.56	15/8 = 1.875	MAJOR 7th
C'	Doh'	TONIC	523.26	2	523.26	2	OCTAVE

Telephone

Basic circuit of a telephone

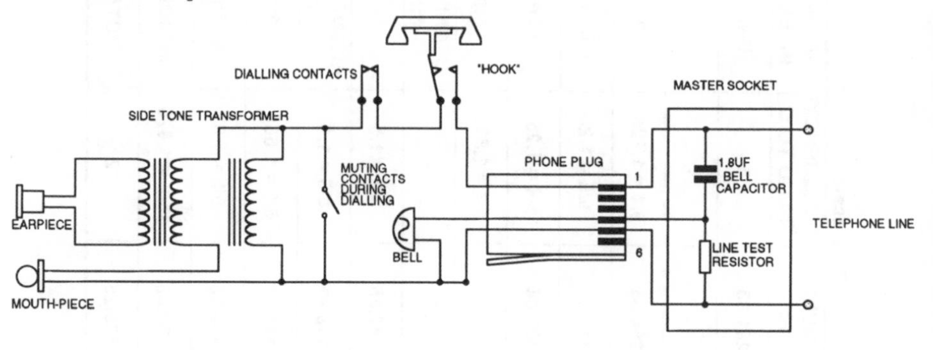

- To ac, the telephone line is a 600 ohm balanced 2-wire transmission line.
- To dc, the telephone line is a 50V dc source with a source resistance of between 400 and 1660 ohms.
- An "on the hook" telephone is open circuit to dc.
- The bell or ringer is ac coupled to the line via a 1.8uF capacitor.
- The ringing voltage is approximately 75V r.m.s. 25Hz (or 16Hz).
- The telephone is "answered" by completing a dc circuit to the telephone line (usually via the mouthpiece).
- The side-tone transformer reduces the mouthpiece's signal level to the earpiece (so the speaker does not deafen himself) but does not attenuate incoming signals.
- "Old fashioned" dialling is called loop-disconnect; the dc path completed by lifting the hand-set is interrupted twice in quick succession for a digit "2", 3 times for a digit "3" and so on; (0 is 10 times).
- Tone dialling is accomplished by simultaneously transmitting two tones for each digit.

DIGITS					LOW FREQUENCY
	1	2	3	A	697Hz
	4	5	6	B	770Hz
	7	8	9	C	852Hz
	*	0	#	D	941Hz
HIGH FREQUENCY	1209Hz	1336Hz	1477Hz	1633Hz	

9 Light

Visible light

Visible light is in the wavelength range **380-760nm**.

Luminous flux

This is the rate of flow of luminous energy and is measured in lumens. One lumen of luminous flux at a wavelength of 400nm corresponds to a radiated power of 3.5 watts. The light output of lamps is specified in lumens.

Illumination

When light hits a solid surface, the process is known as illumination. Illuminance is the luminous flux per unit area and is measured in lux. **1 lux = 1 lumen per square metre**.
In the USA the foot-candle is used where 1 foot-candle = 1 lumen/square foot = 10.76 lux.

APPLICATION	SUGGESTED ILLUMINATION
EMERGENCY LIGHTING ESCAPE ROUTES	0.2 LUX
STORAGE AREAS	1-10 LUX
CAR PARKS, NON-CRITICAL WORK AREAS	10-50 LUX
SPORTS PRACTICE, PLAYGROUNDS	50-100 LUX
FACTORY ASSEMBLY AREAS	300 LUX
GENERAL OFFICE WORK	400-500 LUX
DRAWING OFFICES, INSPECTION AREAS	750 LUX
SPORTING EVENTS FOR TV	500-1000 LUX

Luminous intensity

This is the measure of how much luminous flux is emitted within a small conical angle. Luminous intensity is measured in candelas. A spotlamp concentrates the light output into a narrow beam and so the luminous intensity when looking into the beam is high. A light bulb having the same lumen output would not give the same luminous intensity as the light is radiated over a much greater solid angle.

Luminous efficacy

This is a measure of the light output of a lamp per unit of electrical power consumed. It is measured in lumens per watt.

LAMP TYPE	EFFICACY (LUMENS PER WATT)
GLS (LIGHT BULB)	10 to 20
TUNGSTEN HALOGEN	12 to 22
HIGH PRESSURE MERCURY (MBF)	32 to 56
FLUORESCENT	68 to 80
HIGH PRESSURE SODIUM (SON)	55 to 120
LOW PRESSURE SODIUM (SOX)	70 to 125

Electronics in lighting

An uninformed engineer may be of the opinion that the circuitry involved in lighting is no more complex than a light bulb and a switch. There are many applications of electronics to lighting however; to start and ballast certain light sources such as discharge lamps, to control when the illumination is required, and to control the lamp's intensity.

Control of fluorescent lighting

A fluorescent lamp may not be connected directly to the 240V 50Hz mains supply. The current through the lamp must be limited. The most obvious way would be to use a resistor in series with the lamp, however the resistor would get very hot and the total efficacy (light output per watt of electricity) would be very poor. A capacitor could be used, but this solution would be subject to very large pulses of current through the lamp if the mains power was applied when it was at the peak of the sine-wave. An inductance in series with the lamp is used in practice and is referred to as a ballast. It is simply a coil of wire on a laminated iron core. If the circuit is connected to the mains supply, the lamp will not strike because the gas inside a fluorescent lamp ionizes (and so becomes conductive) at voltages much greater than the 240V mains supply. The lamp must be started and this is achieved by:

1. Heating the cathodes of the lamp to encourage them to emit electrons.
2. Applying a high voltage across the lamp to promote conduction through the gas.

The device used to perform this task is called a starter and it interacts with the ballast to achieve the starting process.

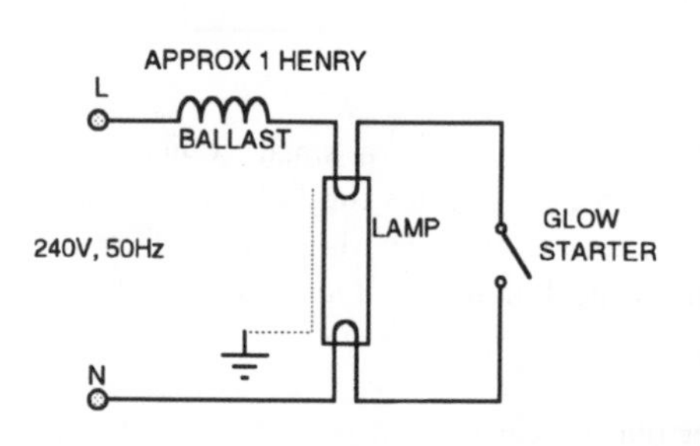

The starter is a bi-metallic strip contact set, encapsulated in a glass envelope which contains a gas. When power is first applied, the bi-metallic contacts are open circuit and a small current flows through the gas causing it to heat up. The bi-metallic contacts close causing a circuit to be connected via the cathodes of the lamp; the current is limited by the reactance of the ballast and the cathodes of the lamp are heated. The gas in the starter cools because the current is now flowing through the contacts. Eventually the contacts open and interrupt the current and because it is flowing in an inductive circuit, a high voltage is generated across the contacts and consequently across the lamp. If the lamp does not strike, then the process repeats until the lamp does conduct. When the lamp is conducting then the voltage across it is much lower than that needed to cause the gas in the starter to heat up, consequently the starter remains open circuit.

Electronic starter

The glow starter is simple and cheap but has a limited life. It does not necessarily start the lamp first time and there may be several attempts accompanied by annoying flashing and flickering of the lamp before conduction is achieved. When the lamp fails, the glow starter makes repeated attempts to start the lamp which results in an incessant flicker. The electronic starter overcomes all of the drawbacks of the glow starter. The lighting industry seems to have of an electronic switch device called the Fluoractor which was pioneered by Texas Instruments.

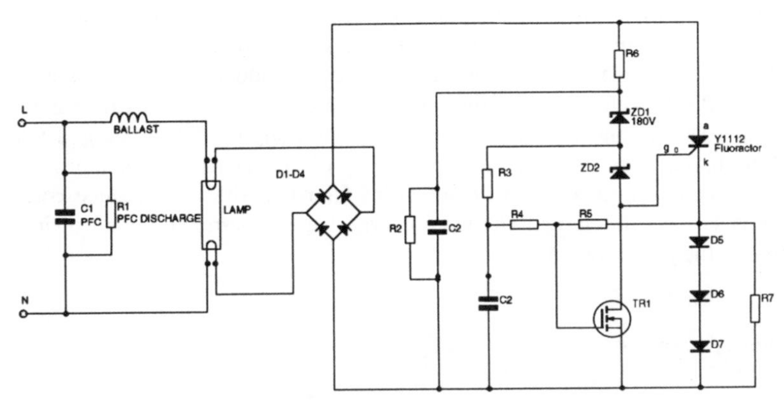

The Fluoractor is effectively a Darlington connected GTO pair of devices in parallel with a 1500V power zener. When power is first applied to the circuit, the Fluoractor device is triggered into conduction via R6, ZD1 and ZD2. Cathode heating of the lamp occurs via the Fluoractor in series with D5, D6 and D7. The voltage pedestal created by the forward voltage drop of D5, D6 and D7 causes C2 to be charged via R4 in series with R5. After a period of a second or so, the gate-source threshold voltage of the FET TR1 is reached, it turns on and causes the Fluoractor to be turned off by GTO action. When the Fluoractor turns off, the current it interrupts is flowing in an inductive circuit (the ballast) and consequently a high voltage back-emf is generated. This is clamped to 1500V by the power zener in the Fluoractor. The process repeats until the lamp strikes. The zener diode, ZD1, is selected to have a voltage which is greater than the running voltage of the lamp, consequently when the lamp is running, ZD1 will not conduct and the Fluoractor is prevented from further triggering.

Electronic ballast

The fluorescent lamp may be ballasted electronically at high frequency and has the following advantages over the wound ballast:

- Higher efficiency.
- Higher efficacy (more light output per watt input).
- Less flicker and a reported lower incidence of headaches of staff working in fluorescent light.
- Less weight than wound ballasts.
- No 50Hz "buzz".

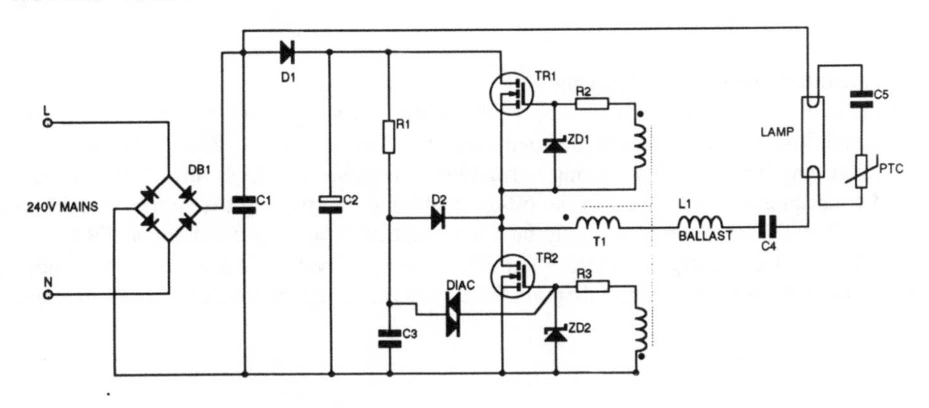

TR1, TR2 and T1 form a self-oscillating inverter running at approximately 30kHz. It is "kick-started" by R1, C3 and the diac. D2 disables the kick-start when the inverter is running. The lamp is ballasted by L1 which is a small-size ferrite-cored inductor. C4 blocks dc to the lamp and C5 and the PTC give cathode heating. C5 resonates with L1 to provide a high voltage to start the lamp. During any one half cycle of the mains supply, C1 is charged via the diode bridge and discharged into the lamp at the frequency of the inverter. This improves the power factor of the circuit considerably. An input mains filter is desirable to reduce conducted r.f.i into the mains supply. Integrated circuits are available to control the input mains current to further improve the power factor.

High frequency inverter

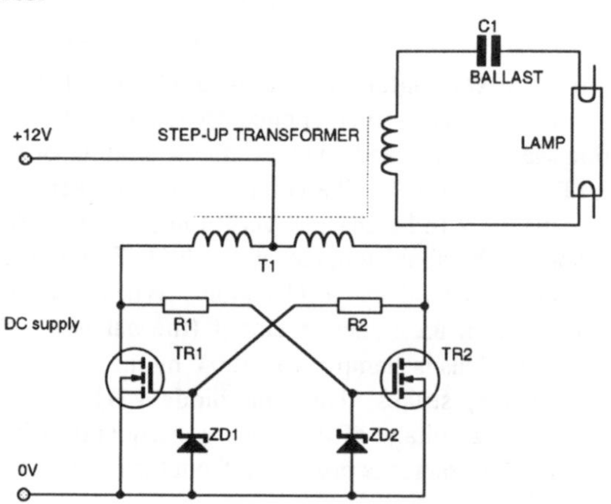

Fluorescent lamps may be operated from a low voltage dc supply such as an automobile battery by means of a simple high frequency inverter circuit. The MOSFETs TR1 and TR2 form an astable multivibrator which operates at approximately 30kHz. The voltage is transformed up by T1, and C1 acts as the current limiting device or ballast. Automotive supplies are notoriously noise-prone, therefore adequate transient protection should be used to prevent damage to the semiconductors. The high frequency inverter may also be used to operate fluorescent lighting from stand-by batteries for emergency lighting applications where mains failure may occur.

Non-maintained emergency lighting

The lamp is only lit when the mains supply falls below a pre-set minimum level. During normal mains conditions, the nickel-cadmium batteries are charged via R4. The LED indicates that battery charging is in progress. During normal mains conditions TR1 is turned on via ZD1 and R2 and consequently TR2 is held in the off-state thereby interrupting the supply to the inverter TR3 and TR4. When the mains supply falls such that ZD1 no longer conducts, TR2 is turned on and the inverter is energized thereby lighting the lamp. Over three hours of emergency light is then provided. A typical circuit diagram for an emergency lighting unit is shown overleaf.

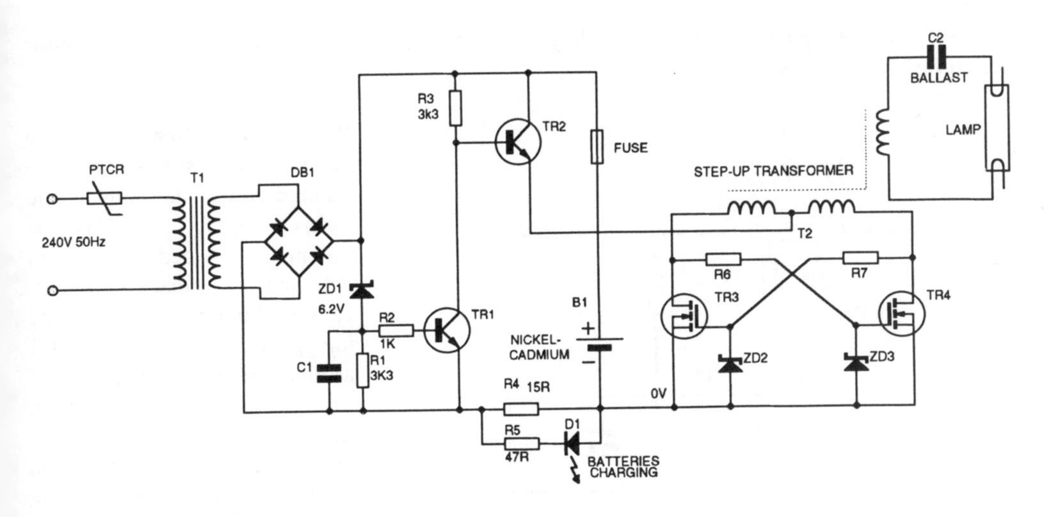

Electronic transformer

The 12V tungsten-halogen lamp may be powered by a 50Hz step-down transformer. However, with lamp wattages of 75W, the transformer starts to become bulky. The electronic transformer is a lightweight solution to supplying power to tungsten-halogen lamps which lends itself to integration with the lamp fitting. The mains supply is full-wave rectified and then inverted at approximately 30kHz. The step-down transformer is then a light-weight, small-size ferrite cored device. The BR100 diac "kick-starts" the inverter at power-up and D1 disables the "kick-start" when the inverter is running.

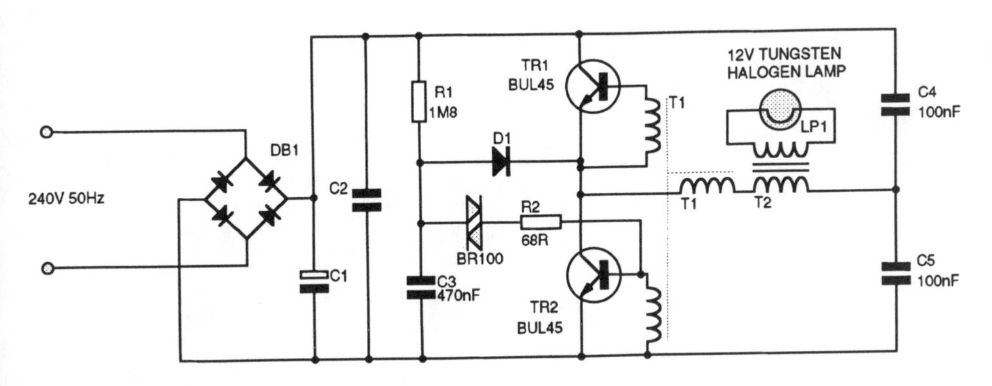

Ignitor

Sodium discharge lamps may not be connected directly to the mains supply as the current through the gas discharge must be limited. This is traditionally accomplished by an inductor called a ballast. The lamp will not start, however, when the mains is connected and a means of ignition is required; this is termed an ignitor. An ignitor is an electronic circuit which interacts with the ballast to produce high voltage pulses in the order of 1.8 to 5kV depending upon the lamp to be ignited. The ballast is usually provided with a tap which enables the ignitor to supply pulses of current through part of the ballast's winding. The ballast acts as a step-up auto-transformer generating the high voltage striking pulses across the lamp. A typical ignitor circuit is shown below.

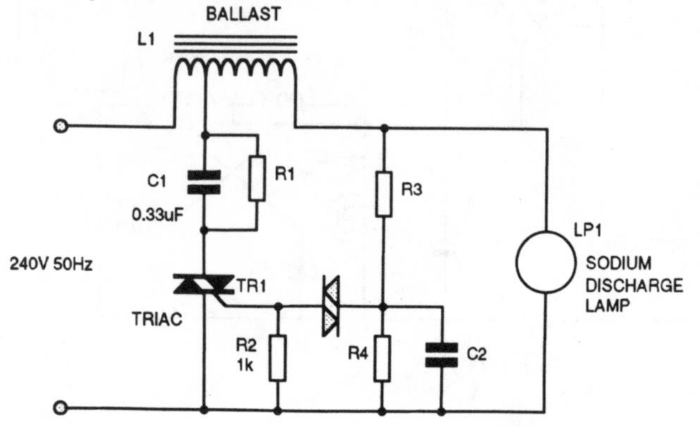

When power is first applied, the lamp is not struck and the voltage across the lamp is the mains supply. This is potentially divided by R3 and R4 and the diac conducts, causing the triac to be triggered into pulsed conduction via C1 around the peak of each half cycle of the mains supply. An ignition pulse is produced each time the triac conducts. When the lamp is struck, the running voltage is lower than the mains supply and the diac ceases to conduct; the ignitor is then disabled.

10 Heat

Temperature coefficients

All electronic components change in value and performance with changes in temperature. The circuit designer must be aware of these changes and allow or compensate for them in the design. In general, semiconductor parameters increase with temperature. This is not always desirable. The gain of a transistor increases with temperature which may improve a circuit's performance but the collector to base leakage current also increases with temperature which can degrade the performance.

Thermal management

Some electronic components generate heat during their operation. If the heat is of a low order, the component's body and connecting leads may be able to dissipate the heat and not cause the component's temperature to rise above the manufacturer's specified maximum limit. If the heat is such that the component's temperature exceeds the maximum then, assuming the design cannot be electronically modified, the heat must be extracted at a faster rate by either improving ventilation or heat-sinking the component.

Appreciation of temperature

The electronics engineer should have a "feel" for temperature levels which occur in electronics and related subjects. The first instinct is to touch the component to estimate its temperature and the "engineer's finger test" is used by many electronics engineers. It should be stressed, however, that extreme caution should be exercised when using the engineer's finger test. The component being touched may exceed 100°C in which case injury may be sustained. Even worse, the component could be live.

The table shown overleaf details some common temperatures which occur in engineering.

ENGINEER'S FINGER TEST (°C)

Skin sticks	-20° or below	Can hold finger on for seconds	60°C
Cool can of beer	4°C	Can hold finger on for 1 sec. *	70°C
Barely getting warm	30°C	Can hold finger on < 1 sec	80°C
Finger temperature	37°C	Finger is withdrawn quickly	+90°C
Getting warm	40°C	Blister (not recommended)	+120°C
Can hold palm on indefinitely	50°C		

* Mnemonic 1 **SE**cond is **SE**venty degrees

Temperature reference chart

CATEGORY	DESCRIPTION	^oC	^oF
Electronics	Maximum silicon junction temperature	150 to 175	302 to 347
	60/40 solder melts	188	370.4
Physics	Water freezes	0	32
	Water boils	100	212
Melting point of metals	Mercury	-39	-38.2
	Sodium	98	208.4
	Tin	232	449.6
	Lead	327	620.6
	Zinc	419	786.2
	Magnesium	659	1218.2
	Aluminium	660	1220.0
	Silver	961	1761.8
	Gold	1063	1945.4
	Copper	1083	1981.4
	Nickel	1455	2651.0
	Iron	1527	2780.6
	Platinum	1773	3223.4
	Tungsten	3387	6128.6
Plastics	PVC max	100	212
	EVA max	140	284
	PTFE Max	250	482
	Silicone rubber	150	302
	Natural rubber	60	140
Domestic	Home freezer	-18	0
	Home fridge	4	39.2
	Feels chilly	10	50
	Comfortable room temperature	25	77
	Feels far too hot	35	95
	Body temperature	36.9	98.4

To convert ^oC to ^oF multiply by 9/5 and add 32
To convert ^oF to ^oC subtract 32 and multiply by 5/9
To convert K to ^oC subtract 273

Temperature coefficients (passive)

Resistors

TYPE	ppm / °C
Carbon composition	+200 to -2000
Carbon film	+200 to -1000
Metal film	±50
Metal oxide	±200
Metal glaze	±200
Wirewound	±100 to ±200
Surface mount	±200

Film capacitors

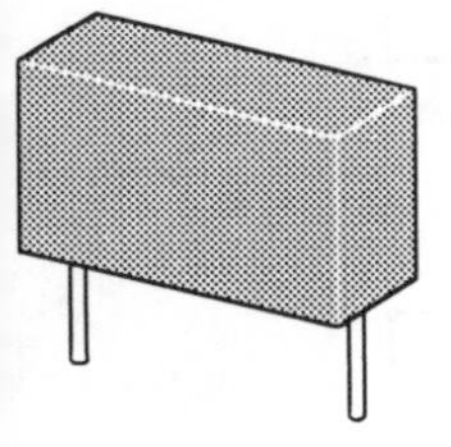

TYPE	ppm / °C (0 to 80°C)
Polyester film / foil	+500
Metallized polyester	+400
Polypropylene film / foil	-200
Metallized polypropylene	-300
Polycarbonate film/foil	-25
Metallized polycarbonate	-25
Polystyrene	-60

Temperature coefficients (active)

Bipolar transistors

All temperature drift effects cause the collector current to rise with temperature.

V_{be}	-2.5mV/ °C
h_{FE}	+2% / °C
I_{CBO}	doubles for every 10°C rise

N channel MOSFETs

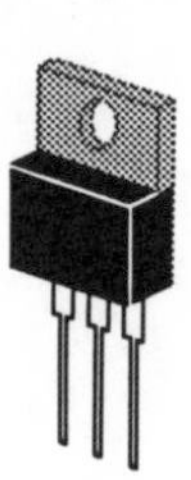

$V_{gs(th)}$	-5mV/ °C
g_{fs}	-0.2% / °C
$R_{DS\ (on)}$	doubles for every 100°C rise
Switching times	Independent of temp

Thyristors and triacs

I_{GT}	Reduces by 10% for every 10°C rise
$I_{holding}$	Reduces by 5% for every 10°C rise
$I_{latching}$	Reduces by 10% for every 10°C rise

Zener diodes (400mW)

ZENER VOLTAGE	TEMP COEFFICIENT
2.7V	-3mV/°C
4.7V	-2mV/°C
5.6V	approx 0mV/°C
9.1V	+5mV/°C
15V	+11mV/°C
30V	+27mV/°C
75V	+80mV/°C

hermal impedance calculations

emiconductors

emiconductor devices are fabricated on a small piece of semiconductor material (such as ilicon) which is commonly called a chip. The chip is mounted inside a package which may be lastic or metal. When the device generates power, heat is produced. The heat flows through e packaging material and if the package is not in thermal contact with another object, the heat ill radiate from the outside of the package to the surroundings. As the heat flows through e package and radiates to the surroundings, a temperature gradient is produced. The chip's emperature is therefore higher than the surroundings by an amount that is proportional to the ower in the chip. The temperature rise of the chip above the surroundings per watt in the chip called the thermal impedance of the package and is given in degrees Celsius or Kelvin per vatt.

An electrical equivalent circuit may be devised to calculate the thermal conditions of the evice. The power in the chip may be represented by a current source. The thermal impedance f the device may be represented by a resistor, and the temperature rise of the chip represented y the voltage produced across the resistor by the current source.

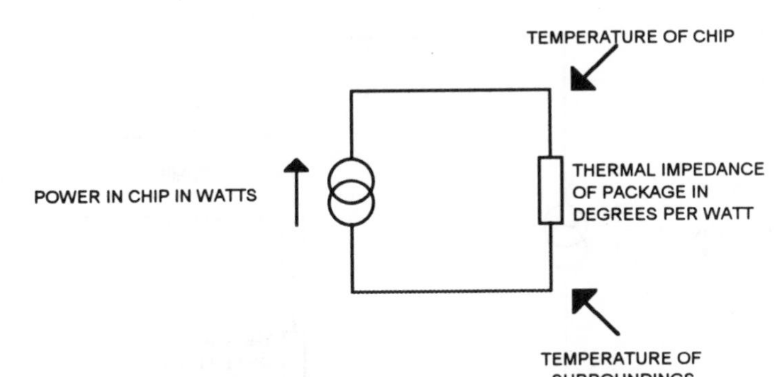

ELECTRICAL EQUIVALENT CIRCUIT TO THE THERMAL CIRCUIT OF A DEVICE

The temperature of the chip is therefore the temperature of the surroundings plus the emperature rise across the thermal impedance of the package. Traditionally, thermal mpedances were quoted for transistors and therefore the term **junction** is used rather than hip.

The symbol used for thermal impedance is the Greek letter theta θ. The symbol representing he total thermal impedance of the package of a device, from its chip (or junction) to the urroundings, is denoted: $\theta_{junc\text{-}ambient}$ or simply $\theta_{j\text{-}a}$.

he temperature of the chip = Ambient temp + P . $\theta_{j\text{-}a}$

Where **P** = power in the chip in watts.)

Heat sinks

The temperature of the chip or junction must not be allowed to exceed a maximum which fo silicon is typically 150 ^{o}C. Manufacturers denote the maxiumum temperature of the chip in device by the symbol T_j. If the calculation shown on the previous page is carried out an shows that the chip's temperature will exceed its maximum T_j then, if it is not possible t reduce the power in the chip or reduce the ambient temperature, the device must be mounte on a heat sink. Heat will flow into the heat sink and will then radiate to the surroundings fror the greater surface area of the heat sink. The heat sink will have a thermal impedance whic quantifies the temperature rise of the heat sink above the surroundings for a given power in th device. This may be given the symbol θ_{HS}.

It may be necessary to electrically insulate the device from the heat sink and the insulato will also have a thermal impedance θ_{INS}. The manufacturer of the device will quote a secon thermal impedance which represents the temperature rise from the junction (chip) to the outsid case of the package. This is given the symbol $\theta_{j\text{-}c}$ ($\theta_{junction\text{-}case}$).

An electrical equivalent circuit may be devised to enable the chip's temperature to b calculated.

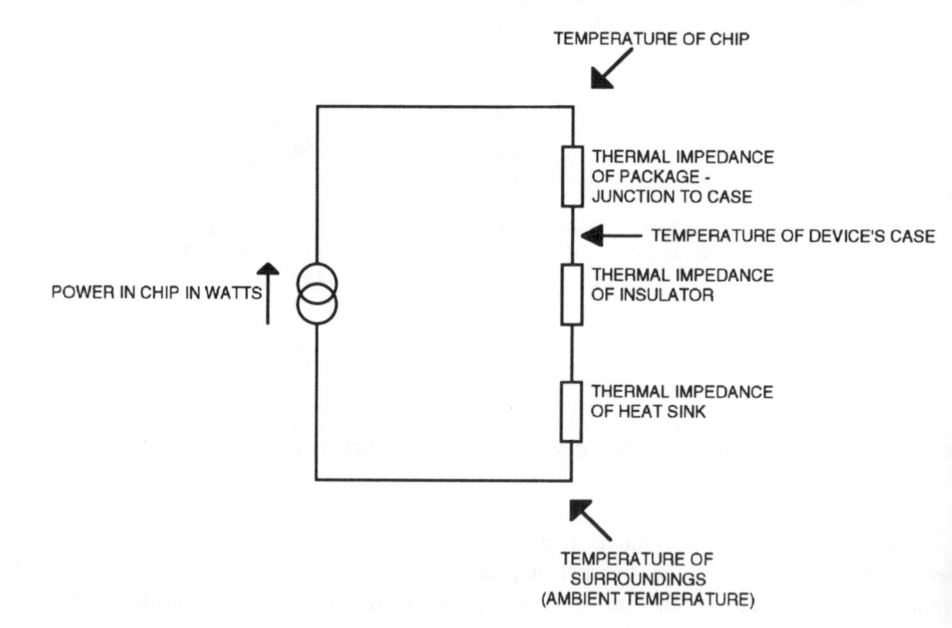

THE ELECTRICAL EQUIVALENT CIRCUIT WHEN THE DEVICE IS CONNECTED TO A HEAT SINK

The temperature of the chip = Ambient temp + P ($\theta_{j\text{-}c}$ + θ_{INS} + θ_{HS})

Thermal impedances of semiconductor packages - plastic

CASE OUTLINE	$\theta_{junc\text{-}free\ air}$ °C /W	$\theta_{junc\text{-}case}$ °C /W	$\theta_{case\text{-}heat\ sink}$ Without insulator °C /W		$\theta_{case\text{-}heat\ sink}$ With insulator °C /W	
			No heat sink compound	With heat sink compound	No heat sink compound	With heat sink compound
TO92	200 to 357	60	-	-	-	-
TO126	83.3 to 100	3.3 to 10				
TO220	62 to 70	1.5 to 4.2	1.2 to 2	0.6 to 1.2		2.1 to 2.6
TO202	62.5 to 75	7.5 to 12.5				
TO218	35.7	1 to 1.56				0.6

Thermal impedances of semiconductor packages - metal

CASE OUTLINE	$\theta_{junc\text{-}free\ air}$ °C /W	$\theta_{junc\text{-}case}$ °C /W	$\theta_{case\text{-}heat\ sink}$ Without insulator °C /W		$\theta_{case\text{-}heat\ sink}$ With insulator °C /W	
			No heat sink compound	With heat sink compound	No heat sink compound	With heat sink compound
TO18	**500**	**150 to 200**				
TO39	**175 to 190**	**25 to 40**				
TO3 (b, e, c)	**32 to 43**	**1**	**0.05 to 0.2**	**0.005 to 0.1**	**0.55 to 0.8**	**0.28 to 0.4**

11 Connections

Integrated circuit pin-outs

Dual in-line integrated circuits

TOP VIEWS

8 7 6 5

HOLD "CUT-OUT" TO LEFT

LM741C

DOT DENOTES PIN 1

1 2 3 4

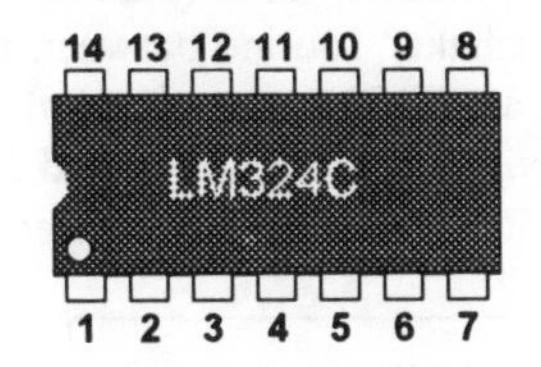

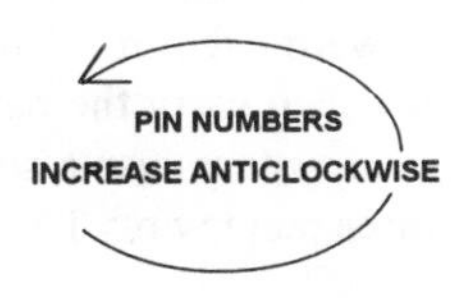

Metal package

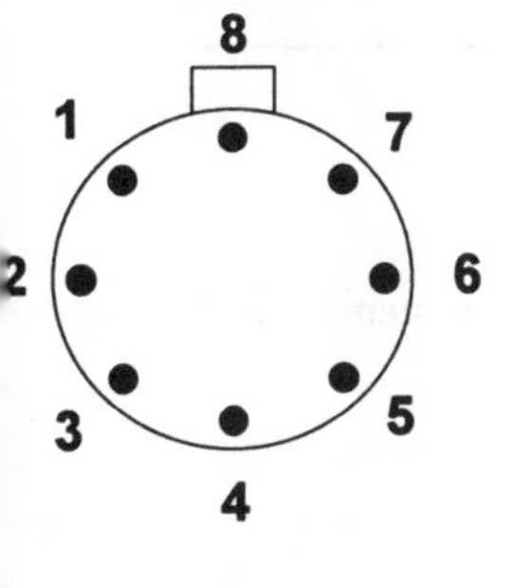

- Highest numbered pin is always next to tab.
- Pins are numbered anticlockwise.
- A metal package may be inserted into a dual in-line position by arranging the pins into two rows.

Single in-line integrated circuits

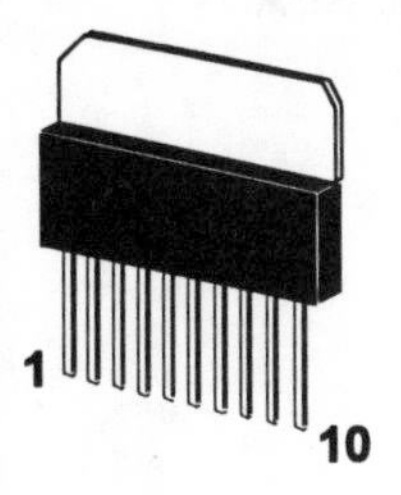

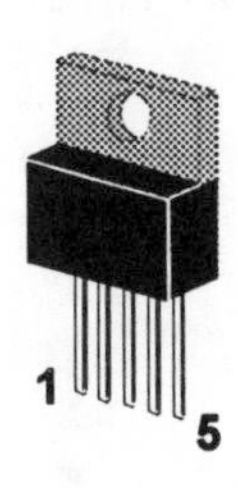

The pins are numbered from left to right looking at the IC legend.

RS232C

RS232C is a serial communication system or standard which was originally developed to enabl computer terminals to communicate with modems. It has gradually evolved into a commo serial communication system between computers and peripherals such as printers (or even othe computers).

The original system used nine or more wires between the computer terminal DTE (dat terminal equipment) and the modem DCE (data circuit termination equipment). Many of th wires were used in a "handshake" routine between the computer and the modem. The mai wires which carry the data are TXD (transmit data), RXD (receive data) and common or 0V Modern systems use a software "handshake" called Xon Xoff protocol and in these cases onl the three main wires TXD, RXD and 0V are required between the two items of equipment.

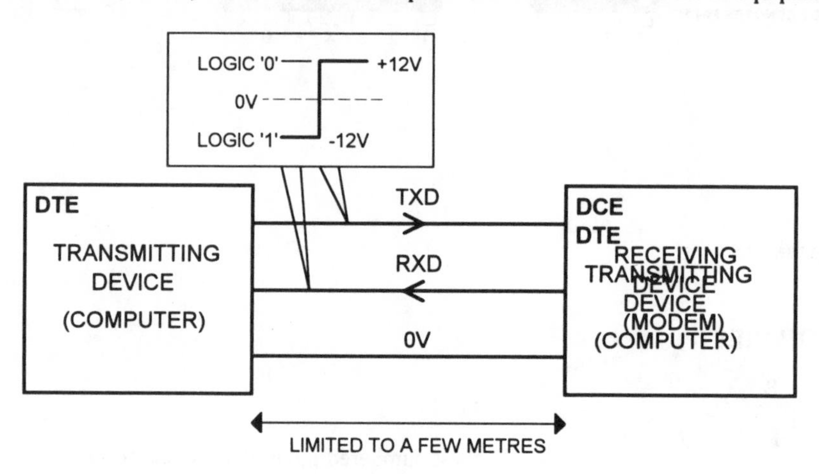

The most popular type of connector which is used for the RS232C system is the 25 way 'D' connector.

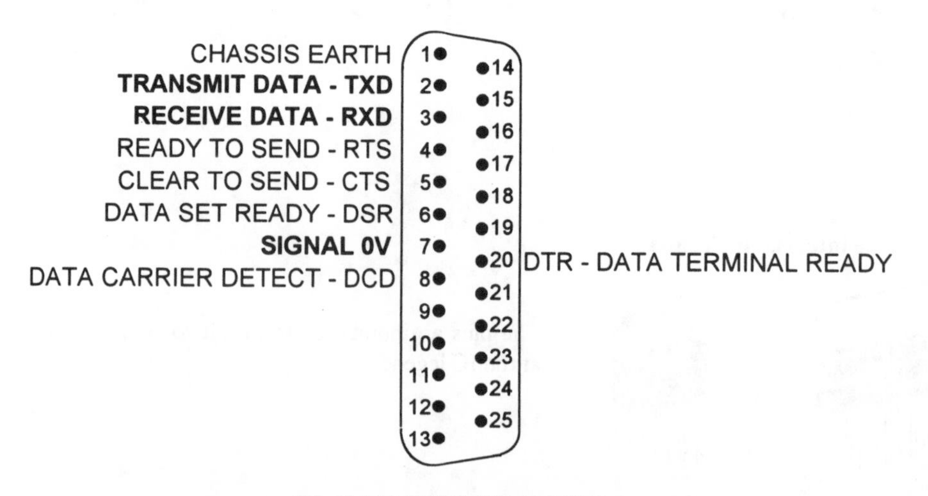

25 WAY "D" CONNECTOR
LOOKING AT SOLDER CONNECTIONS

Making an RS232C connection between computer and peripheral

If there is software Xon Xoff protocol used by the equipment, then the simple connection below may work.

COMPUTER PERIPHERAL

LOOKING AT SOLDER CONNECTIONS OF PLUGS

If the above does not work, the "hardware handshake" at the computer may have to be defeated as below.

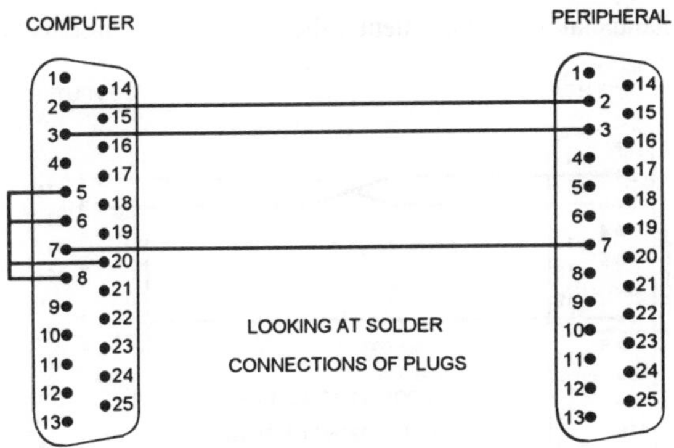

If the above does not work, the peripheral may require a DTR and RTS signal as below.

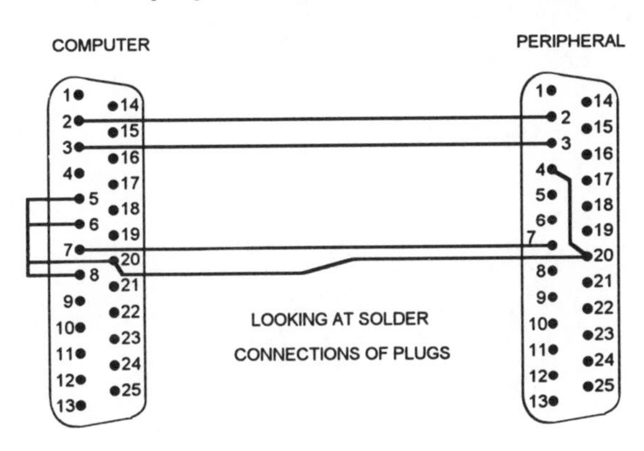

Making an RS232C connection between computers

When two computers are connected, the TXD and RXD connections have to be mirrored. Some peripherals may also need this mirrored connection. If there are no hardware handshaking requirements to satisfy, the connection arrangement below may work.

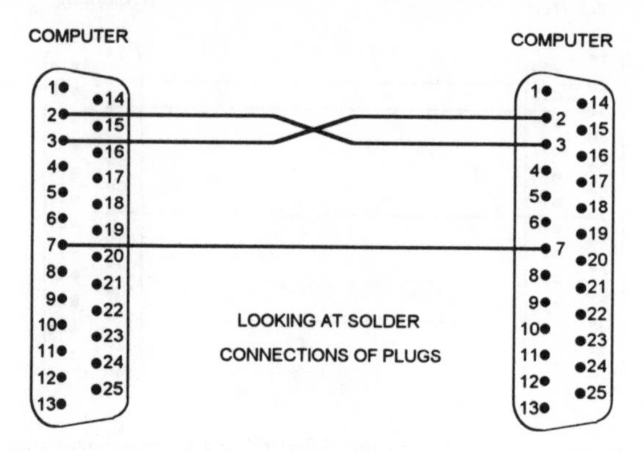

If there are hardware handshaking requirements, they may be defeated as below.

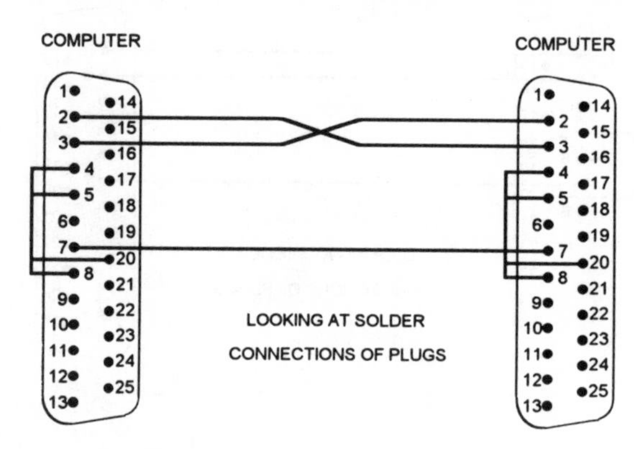

The RS232C system is such that if an error is made during interconnection, no damage will occur to the equipment.

Audio connections

5 pin DIN

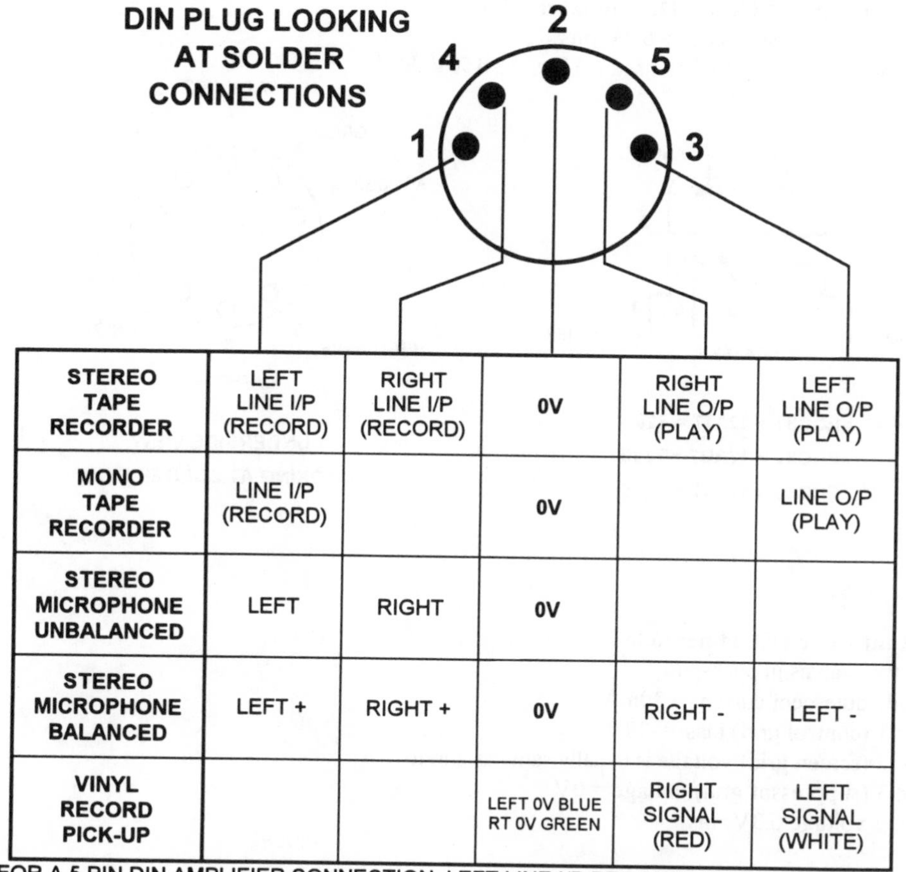

STEREO TAPE RECORDER	LEFT LINE I/P (RECORD)	RIGHT LINE I/P (RECORD)	0V	RIGHT LINE O/P (PLAY)	LEFT LINE O/P (PLAY)
MONO TAPE RECORDER	LINE I/P (RECORD)		0V		LINE O/P (PLAY)
STEREO MICROPHONE UNBALANCED	LEFT	RIGHT	0V		
STEREO MICROPHONE BALANCED	LEFT +	RIGHT +	0V	RIGHT -	LEFT -
VINYL RECORD PICK-UP			LEFT 0V BLUE RT 0V GREEN	RIGHT SIGNAL (RED)	LEFT SIGNAL (WHITE)

FOR A 5 PIN DIN AMPLIFIER CONNECTION, LEFT LINE I/P BECOMES LEFT LINE O/P ETC.

3 pin XLR connector

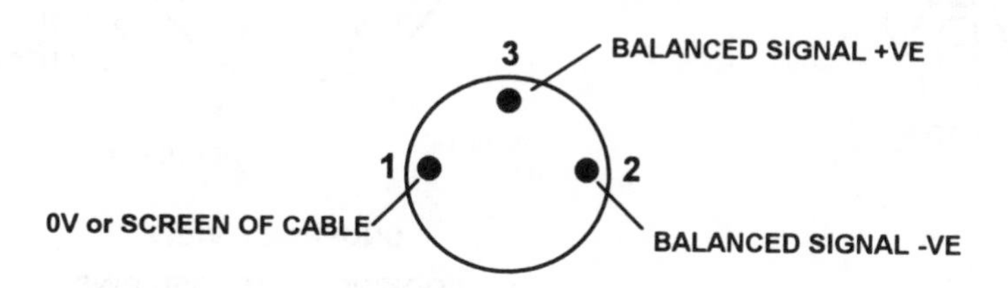

Thermionic valve pin-outs

Thermionic valves are still used today in musical instrument amplifiers as manufactured by companies such as Marshall. There are a limited number of types used today and they are shown below.

Pre-amp valves ECC81, 82, 83(dual triode)
Heater voltage (each side) = 6.3V rms
Typical anode voltage when used in class A =150V dc

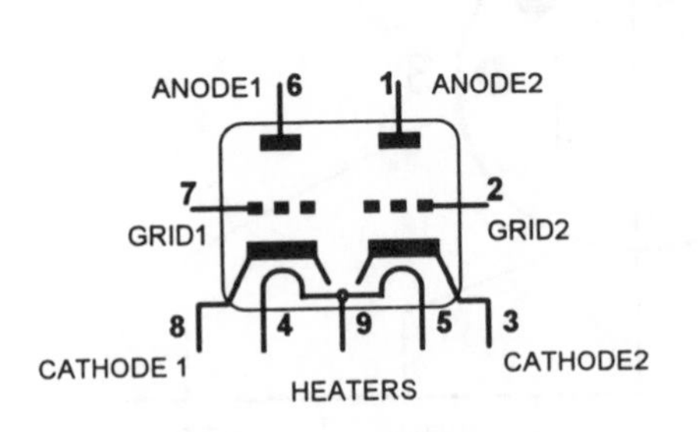

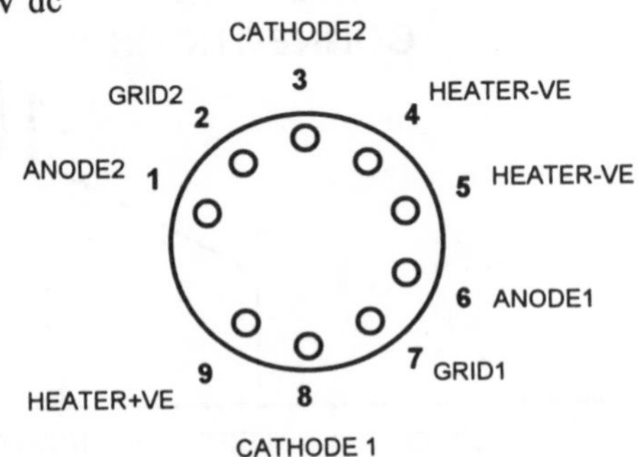

ECC81 = 12AT7 = 6201

ECC82 = 12AU7 = 7730

ECC83 = 12AX7 = 7025

UNDERSIDE VIEW

LOOKING AT SOLDER PINS

Output valve, EL34 pentode
For two valves in push-pull
Anode quiescent current = 30mA
Grid 1 (control grid) bias = -33V
Grid 2 (screen grid) voltage is usually same as anode
Grid 3 (suppressor grid) voltage = 0V
Heater voltage 6.3V rms

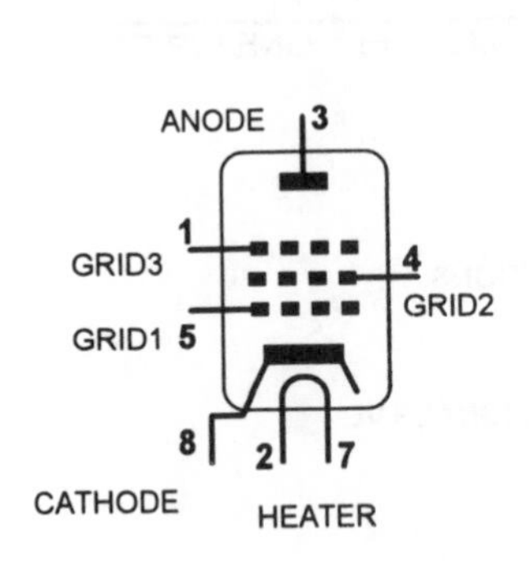

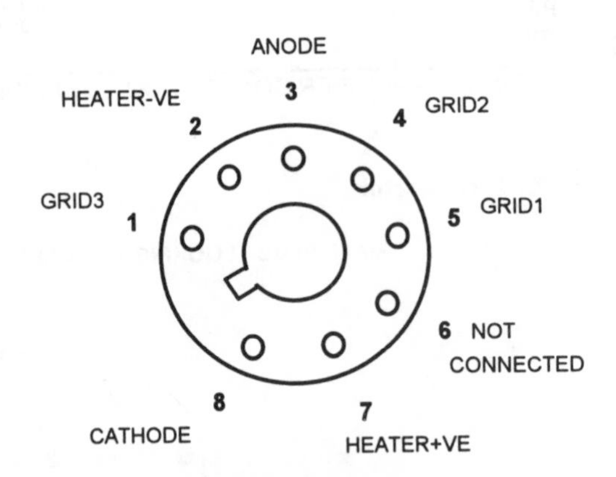

EL34

UNDERSIDE VIEW

LOOKING AT SOLDER PINS

Output valve, EL84 pentode
For two valves in push-pull
Anode quiescent current = 30mA
Grid 2 (screen grid) voltage is usually same as anode
Grid 3 connected to cathode
Heater voltage 6.3V rms (760mA)

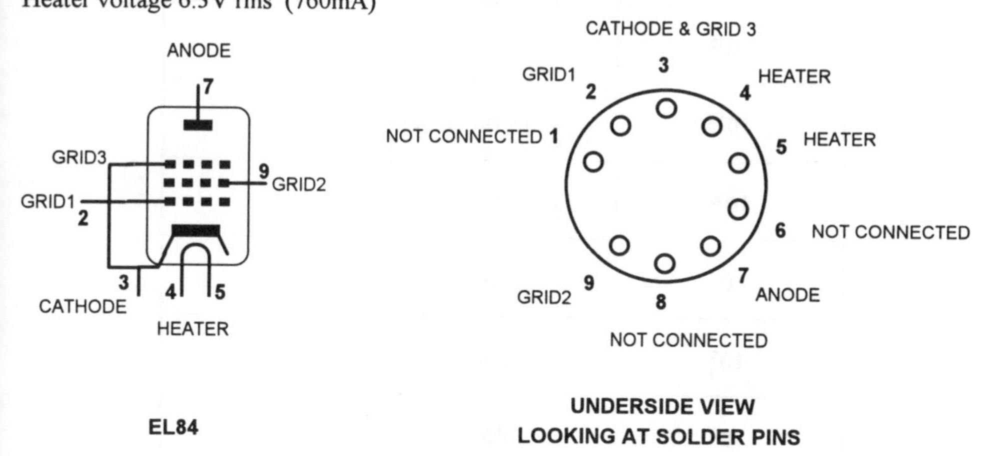

21 pin SCART connector

All designations are with respect to the appliance the plug is inserted into e.g. If plugged into a TV the "audio output" means an audio signal from the TV.

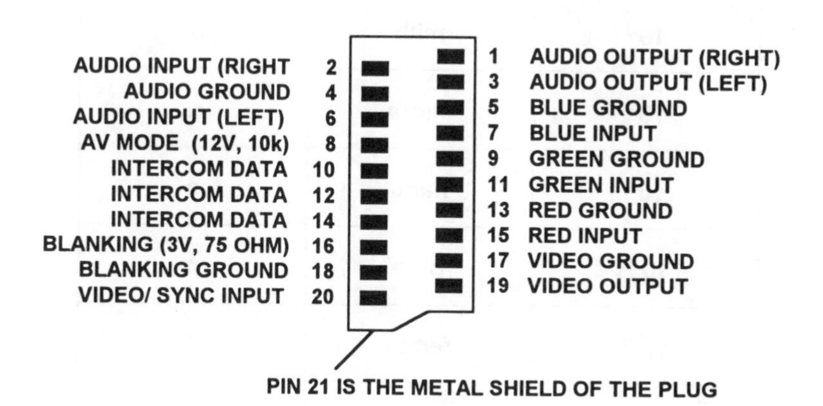

Appendix

Decimal prefixes

POWER OF 10	PREFIX	PREFIX LETTER
10^{18}	exa	E
10^{15}	peta	P
10^{12}	tera	T
10^{9}	giga	G
10^{6}	mega	M
10^{3}	kilo	k
10^{2}	hecto	h
10	deca	da
10^{-1}	deci	d
10^{-2}	centi	c
10^{-3}	milli	m
10^{-6}	micro	μ
10^{-9}	nano	n
10^{-12}	pico	p
10^{-15}	femto	f
10^{-18}	atto	a

Greek alphabet

The Greek alphabet has 24 letters and some are used to denote certain parameters in electronic engineering.

NAME	LOWER CASE	DESIGNATION OF LOWER CASE	UPPER CASE	DESIGNATION OF UPPER CASE
ALPHA	α	COMMON BASE CURRENT GAIN	A	
BETA	β	COMMON EMITTER CURR GAIN	B	
GAMMA	γ		Γ	
DELTA	δ	PARTIAL DERIVATIVE	Δ	A SMALL CHANGE
EPSILON	ε	PERMITTIVITY	E	
ZETA	ζ	DAMPING RATIO	Z	
ETA	η	EFFICIENCY	H	
THETA	θ	ANGLE	Θ	
IOTA	ι		I	
KAPPA	κ		K	
LAMBDA	λ	WAVELENGTH, FLUX LINKAGE	Λ	
MU	μ	1 MILLIONTH, PERMEABILITY	M	
NU	ν		N	
XI	ξ		Ξ	
OMIKRON	ο		O	
PI	π	RATIO CIRCUMF TO DIAMETER	Π	
RHO	ρ	RESISTIVITY, CHARGE DENSITY	P	
SIGMA	σ	CONDUCTIVITY	Σ	THE SUM OF...
TAU	τ	TIME CONSTANT	T	
UPSILON	υ		Y	
PHI	φ	PHASE ANGLE, MAGNETIC FLUX	Φ	
CHI	χ		X	
PSI	ψ	ANGLE, ELECTRIC FLUX	Ψ	
OMEGA	ω	ANGULAR FREQUENCY RADIANS/SEC	Ω	OHMS

Delta-star, star-delta transformations

Delta-star transformation

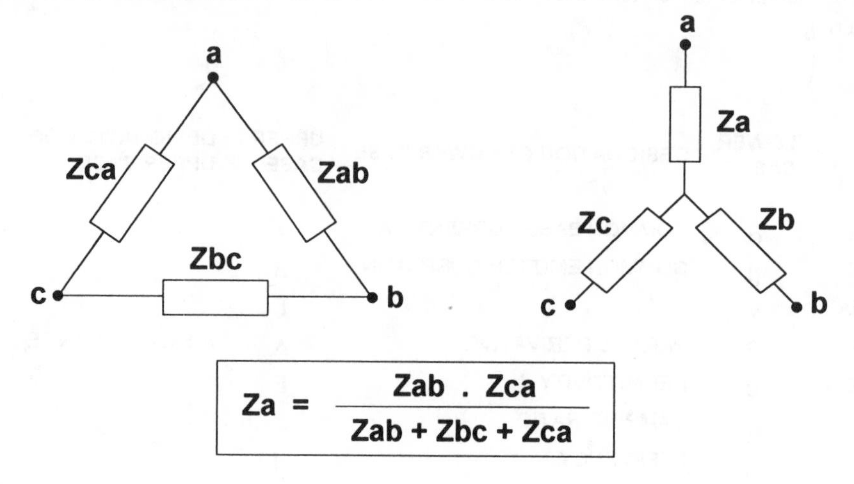

The star impedance is equal to the product of the two impedances at the node divided by the sum of the impedances.

Star-delta transformation

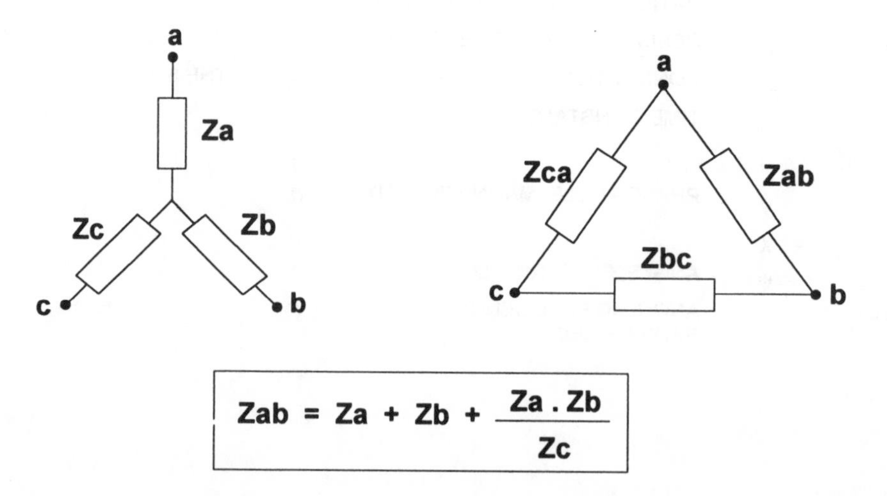

Add the two impedances between the points then add their product divided by the third impedance.

Physical constants

CONSTANT	SYMBOL	VALUE	UNITS
Avogadro's No	N	6.022×10^{26}	atoms /kg mole
Acceleration of gravity	g	9.807	m/s^2
Boltzmann's constant	k	1.38×10^{-23}	J/K
Characteristic Z of free space	$Z_0 = (\mu_o / \varepsilon_o)^{-1/2}$	120π	Ohms
Electron-volt	eV	1.602×10^{-19}	J
Electron charge	e	1.602×10^{-19}	Coulombs
Electron rest mass	m_e	9.109×10^{-31}	kg
Electron charge / mass ratio		1.759×10^{11}	C/kg
Energy for T = 290K	kT	4×10^{-21}	J
Energy of ground state atom		13.6	eV
Faraday Constant	F	9.65×10^{7}	C / kg - mole
Mechanical equivalent of heat		4.1855	Joules/cal
Permeability of free space	μ_o	$4\pi \times 10^{-7}$	H/m
Permittivity of free space	ε_o	$1/36\pi \times 10^{-9}$	F/m
Planck's constant	h	6.626×10^{-34}	Joule - secs
Proton mass	m_p	1.672×10^{-27}	kg
Universal gas constant	R	8.314	kJ/kg mole °K
Velocity of light in a vacuum	c	2.998×10^{8}	m/s
Velocity of sound in air @ sea lvl		1100	ft/s
Volume of 1kg mole of gas@NTP		22.42	m^3

UNITS CONVERSION CHART - LENGTH

ANG-STROM	MICRO-METRE	MILLI-METRE	CENTI-METRE	INCH	FOOT (ft)	YARD (yds)	METRE (m)	FATHOM	KILO-METRE	MILE	UK NAUT. MILE
1	0.0001	1.000E-07	1.000E-08	3.937E-09	3.281E-10	1.094E-10	1.00E-10	5.468E-11	1.000E-13	6.214E-14	5.397E-14
10000	1	1.000E-03	1.000E-04	3.937E-05	3.281E-06	1.094E-06	1.00E-06	5.468E-07	1.000E-09	6.214E-10	5.397E-10
1.000E+07	1000	1	0.1	3.937E-02	3.281E-03	1.094E-03	0.0010	5.468E-04	1.000E-06	6.214E-07	5.397E-07
1.000E+08	1.000E+04	10	1	0.3937	3.281E-02	1.094E-02	0.0100	5.468E-03	1.000E-05	6.214E-06	5.397E-06
2.540E+08	2.540E+04	25.4	2.54	1	8.333E-02	2.778E-02	0.0254	1.389E-02	2.540E-05	1.578E-05	1.371E-05
3.048E+09	3.048E+05	304.8	30.48	12	1	0.3333	0.3048	0.1667	3.048E-04	1.894E-04	1.645E-04
9.144E+09	9.144E+05	914.4	91.44	36	3	1	0.9144	0.5	9.144E-04	5.682E-04	4.935E-04
1.000E+10	1.000E+06	1.000E+03	100	39.37	3.281	1.094	1	0.5468	1.000E-03	6.214E-04	5.397E-04
1.829E+10	1.829E+06	1.829E+03	182.9	72.01	6.001	2	1.829	1	1.829E-03	1.136E-03	9.869E-04
1.000E+13	1.000E+09	1.000E+06	100000	39370	3281	1094	1000	546.8	1	0.6214	0.5397
1.609E+13	1.609E+09	1.609E+06	160934	63360	5280	1760	1609	880.0	1.609	1	0.8685
1.853E+13	1.853E+09	1.853E+06	185318	72960	6080	2027	1853	1013	1.853	1.152	1

UNITS CONVERSION CHART - VOLUME

CUBIC mm	MILLI-LITRES	CUBIC INCHES	UK FLUID OUNCE	US FLUID OUNCE	US PINT	UK PINT	LITRE	US GALLON	UK GALLON	CUBIC FEET	CUBIC YARDS	CUBIC METRES
1	0.001	6.102E-05	3.520E-05	3.381E-05	2.113E-06	1.760E-06	1.000E-06	2.642E-07	2.200E-07	3.531E-08	1.308E-09	1.000E-09
1000	1	6.102E-02	3.520E-02	3.381E-02	2.113E-03	1.760E-03	1.000E-03	2.642E-04	2.200E-04	3.531E-05	1.308E-06	1.000E-06
16387	16.387	1	0.5767	0.5541	3.463E-02	2.884E-02	1.639E-02	4.329E-03	3.605E-03	5.787E-04	2.143E-05	1.639E-05
2.841E+04	28.41	1.734	1	0.9607	6.004E-02	4.999E-02	2.841E-02	7.505E-03	6.249E-03	1.003E-03	3.716E-05	2.841E-05
2.957E+04	29.57	1.804	1.041	1	6.249E-02	5.204E-02	2.957E-02	7.812E-03	6.504E-03	1.044E-03	3.868E-05	2.957E-05
4.732E+05	473.18	28.88	16.65	16	1	0.8327	0.4732	0.1250	0.1041	1.671E-02	6.189E-04	4.732E-04
5.683E+05	568.3	34.68	20	19.22	1.201	1	0.5683	0.1501	0.1250	2.007E-02	7.433E-04	5.683E-04
1.000E+06	1000	61.02	35.20	33.81	2.113	1.760	1	0.2642	0.2200	3.531E-02	1.308E-03	1.000E-03
3.785E+06	3785.4	231.0	133.2	128	8	6.661	3.785	1	0.8327	0.1337	4.951E-03	3.785E-03
4.546E+06	4546.1	277.4	160	154	9.608	8	4.546	1.201	1	0.1605	5.946E-03	4.546E-03
2.832E+07	28320	1728	996.7	958	59.85	49.84	28.32	7.481	6.230	1	3.704E-02	2.832E-02
7.646E+08	764559	46656	26909	25853	1616	1345	764.6	202	168.2	27	1	0.7646
1.000E+09	1.000E+06	61024	35195	33814	2113	1760	1000	264.2	220	35.31	1.308	1

UNITS CONVERSION CHART - VELOCITY

FEET PER MINUTE	CM/SEC	METRES/ MINUTE	INCHES/ SECOND	KMETRES/ HOUR	FEET/ SECOND	MILES/ HOUR	METRES/ SECOND
1	0.50800001	0.3048	0.2	1.829E-02	1.667E-02	1.136E-02	5.08E-03
1.969	1	0.6	0.3937	3.600E-02	3.281E-02	2.237E-02	0.01
3.281	1.667	1	0.6562	0.06	5.468E-02	3.728E-02	1.67E-02
5	2.54	1.524E+00	1	9.144E-02	8.333E-02	5.682E-02	2.54E-02
54.68	27.78	16.67	10.94	1	0.9113	0.6214	0.2778
60	30.48	18.29	12	1.097	1	0.6818	0.3048
88	44.70	26.82	17.60	1.609	1.467	1	0.447
196.9	100	60	39.37	3.6	3.281	2.237	1

UNITS CONVERSION CHART - MASS

MICRO-GRAM	MILLI-GRAM	GRAM	OUNCE	POUND	KILO-GRAM	STONE	SLUG	HUNDRED WEIGHT	METRIC TONNE	UK TON
1	0.001	1.000E-06	3.527E-08	2.205E-09	1.000E-09	1.575E-10	6.853E-11	1.968E-11	1.000E-12	9.854E-13
1000	1	1.000E-03	3.527E-05	2.205E-06	1.000E-06	1.575E-07	6.853E-08	1.968E-08	1.000E-09	9.854E-10
1.000E+06	1000	1	3.527E-02	2.205E-03	1.000E-03	1.575E-04	6.853E-05	1.968E-05	1.000E-06	9.854E-07
2.835E+07	2.835E+04	2.835E+01	1	6.250E-02	2.835E-02	4.464E-03	1.943E-03	5.580E-04	2.836E-05	2.794E-05
4.536E+08	4.536E+05	4.536E+02	1.600E+01	1	4.536E-01	7.143E-02	3.108E-02	8.929E-03	4.537E-04	4.470E-04
1.000E+09	1.000E+06	1.000E+03	3.527E+01	2.205E+00	1	1.575E-01	6.853E-02	1.968E-02	1.000E-03	9.854E-04
6.350E+09	6.350E+06	6.350E+03	2.240E+02	1.400E+01	6.350E+00	1	4.352E-01	1.250E-01	6.352E-03	6.258E-03
1.466E+10	1.466E+07	1.466E+04	5.171E+02	3.232E+01	1.466E+01	2.309E+00	1	2.886E-01	1.466E-02	1.445E-02
5.080E+10	5.080E+07	5.080E+04	1.792E+03	1.120E+02	5.080E+01	8.000E+00	3.481E+00	1	5.081E-02	5.006E-02
1.000E+12	1.000E+09	1.000E+06	3.527E+04	2.205E+03	1.000E+03	1.575E+02	6.853E+01	1.968E+01	1	9.852E-01
1.016E+12	1.016E+09	1.016E+06	3.584E+04	2.240E+03	1.016E+03	1.600E+02	6.963E+01	2.000E+01	1.016E+00	1

UNITS CONVERSION CHART - PRESSURE

mm Hg-TORR	kPASCAL - kN/m sq	cm Hg	ft water	inches Hg	ibf / sq in	kgf/sq cm	BAR	ATMOS - PHERE
1	0.1333	0.1	4.461E-02	3.937E-02	1.934E-02	1.360E-03	1.333E-03	1.316E-03
7.5	1	0.75	0.3346	0.2953	0.145	1.020E-02	1.000E-02	9.868E-03
10	1.333	1	0.4461	0.3937	0.1934	1.360E-02	1.333E-02	1.316E-02
22.42	2.989	2.242	1	0.8826	0.4335	3.048E-02	2.989E-02	2.950E-02
25.4	3.387	2.54	1.133	1	0.4912	3.453E-02	3.387E-02	3.342E-02
51.71	6.895	5.171	2.307	2.036	1	7.031E-02	6.895E-02	6.804E-02
735.5	98.0665	73.55	32.81	28.96	14.22	1	0.9807	9.678E-01
750	100	75	33.46	29.53	14.5	1.02	1	0.9868
760	101.3	76	33.9	29.92	14.7	1.033	1.013E+00	1

UNITS CONVERSION CHART - ENERGY, WORK, HEAT

ELECTRON VOLT	ERG	JOULE (Nm, Wsec)	FOOT POUND	CALORIE	B.T.U.s	KILO-CALORIE	MEGA JOULE	HORSE POWER-HOUR	kW - HOUR kWh	THERM British Gas
1	1.602E-12	1.602E-19	1.182E-19	3.826E-20	1.518E-22	3.826E-23	1.602E-25	5.966E-26	4.450E-26	1.518E-27
6.242E+11	1	1.000E-07	7.376E-08	2.388E-08	9.478E-11	2.388E-11	1.000E-13	3.724E-14	2.778E-14	9.478E-16
6.242E+18	1.000E+07	1	0.7376	2.388E-01	9.478E-04	2.388E-04	1.000E-06	3.724E-07	2.778E-07	9.478E-09
8.463E+18	1.356E+07	1.356E+00	1	0.3238	1.285E-03	3.238E-04	1.356E-06	5.049E-07	3.766E-07	1.285E-08
2.614E+19	41869998.6	4.187	3.088	1	3.968E-03	1.000E-03	4.187E-06	1.559E-06	1.163E-06	3.968E-08
6.586E+21	1.0551E+10	1.055E+03	7.783E+02	2.520E+02	1	0.2520	1.055E-03	3.930E-04	2.931E-04	1.000E-05
2.614E+22	4.187E+10	4.187E+03	3.088E+03	1.000E+03	3.968	1	4.187E-03	1.559E-03	1.163E-03	3.968E-05
6.242E+24	1E+13	1.000E+06	7.376E+05	2.388E+05	947.8	238.8	1	0.3724	0.2778	9.478E-03
1.676E+25	2.685E+13	2.685E+06	1.981E+06	6.413E+05	2.545E+03	641.3	2.685	1	0.7458	2.545E-02
2.247E+25	3.600E+13	3.600E+06	2.655E+06	8.598E+05	3.412E+03	859.8	3.6	1.341	1	3.412E-02
6.586E+26	1.0551E+15	1.055E+08	7.783E+07	2.520E+07	1.000E+05	2.520E+04	105.5	39.30	29.31	1

Tolerance and error calculations

Resistors in series or parallel

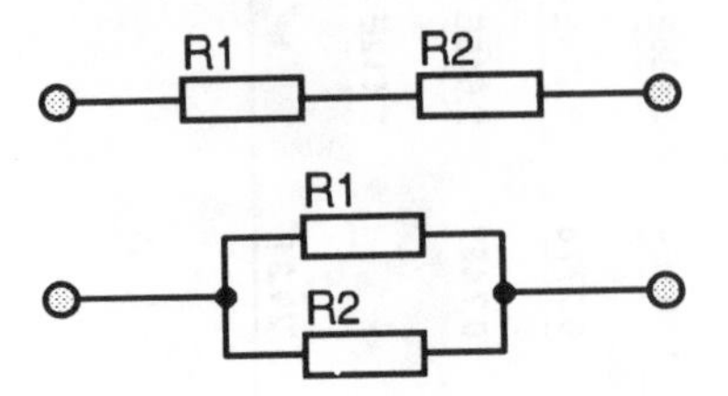

r_1 = % tolerance in R_1	
r_2 = % tolerance in R_2	and $r_1 = r_2 = r$

When the resistors have equal % tolerance then the % tolerance in the total resistance = r

Potential divider

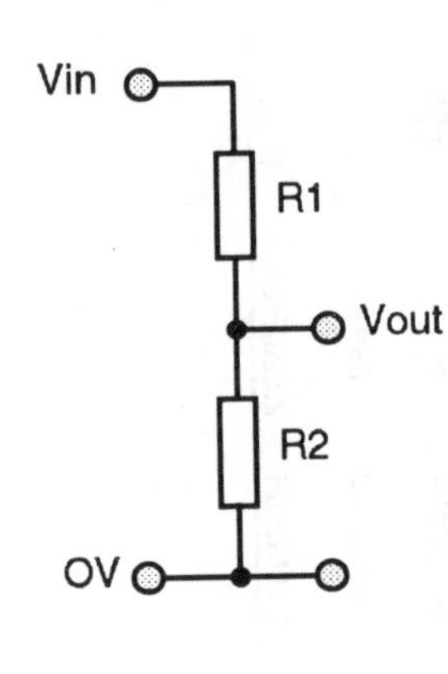

When there is an error in R_1 only, then:

the % error in $V_{out} / V_{in} \cong r_1 \; R_1 / (R_1 + R_2)$

When there is an error in R_2 only, then:

the % error in $V_{out} / V_{in} \cong r_2 \; R_1 / (R_1 + R_2)$

When there is an error in R_1 and R_2 ; and $r_2 = - r_1 = r$ then:

the worst case % error in $V_{out} / V_{in} = r$

Inverting Amplifier

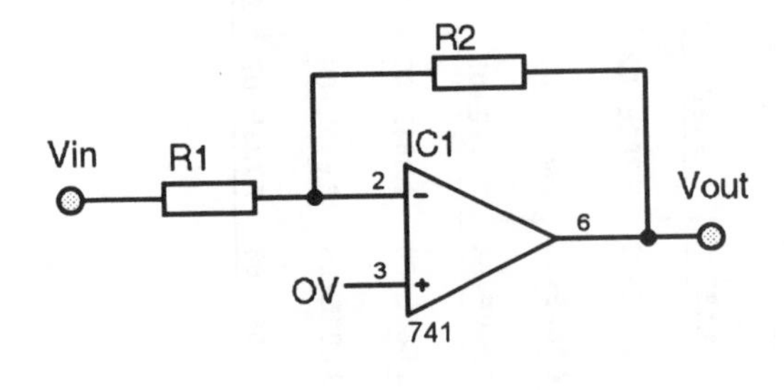

$V_{out} / V_{in} = - R_2 / R_1$

r_1 = % error in R_1

r_2 = % error in R_2

The % error in V_{out} / V_{in}

(1st aprx) = $r_1 - r_2 - r_1 r_2 / 100$

(2nd aprx) = $(r_1 - r_2 - r_1 r_2 / 100)(1 + r_2{}^2 / 10^4) + r_2{}^2 / 10^2$

(3rd aprx) = $(r_1 - r_2 - r_1 r_2 / 100)(1 + r_2{}^2 / 10^4 + r_2{}^4 / 10^8) + r_2{}^2 / 10^2 + r_2{}^4 / 10^6$

Bibliography

Title	Author	Publisher
A Dictionary of Electronics	S Handel	Penguin Books
Electrical Engineering Circuits	Hugh Hildreth Skilling	J Wiley & Sons
Electronic Circuits Handbook	M Tooley	Butterworth-Heinemann
Electronic Designer's Handbook	T K Hemingway	Business Publications
Electronics Pocket Book	E A Parr	Butterworth-Heinemann
The Art of Electronics	Horowitz Hill	Cambridge
Thorn Lighting - Technical Handbook		Editor: G Williams

Technical data books from the following companies:
Arcotronics
Draloric
ERG
MFR
Neohm
Piher
Motorola
Philips Components
Rifa-Evox
Roederstein
SGS - Thomson
Siemens
Texas Instruments
VTM
Welwyn
Wima

Index